PROBLEMS IN MATHEMATICAL ANALYSIS

PROBLEMY MATEMATICHESKOGO ANALIZA

ПРОБЛЕМЫ МАТЕМАТИЧЕСКОГО АНАЛИЗА

PROBLEMS IN MATHEMATICAL ANALYSIS

Series editor: Academician V. I. Smirnov

Volume 1: BOUNDARY VALUE PROBLEMS AND INTEGRAL EQUATIONS — 1968

On the Theory of Simple and Double-Layer Potentials for Domains with Irregular Boundaries • Yu. D. Burago, V. G. Maz'ya, and V. D. Sapozhnikova

Solution of the Third Boundary Value Problem by Potential Theory Methods for Domains with Irregular Boundaries • V. D. Sapozhnikova

On the Modulus of Continuity of the Solution of the Dirichlet Problem near an Irregular Boundary • V. G. Maz'ya

On Cosserat Functions • S. G. Mikhlin

On the Stability of the Index of a One-Dimensional Singular Operator with Vanishing Symbol • Z. Presdorf

Cauchy Problem for Some Degenerating Fourth-Order Hyperbolic Equations • M. M. Smirnov

Dirichlet Problem for Nonuniform Elliptic Quasilinear Equations • T. B. Solomyak

On an Estimate of the Rate of Convergence of Homogeneous Difference Schemes for Elliptic and Parabolic Equations with Discontinuous Coefficients • V. Ya. Rivkind

Continuous Dependence of the Maximum Index of Exponential Growth of the Solution of a Linear Hamiltonian Equation with Periodic Operator Coefficients • V. I. Derguzov

Dynamic Instability Domains of Parametrically Excited Systems with an Infinite Number of Degrees of Freedom • V. N. Fomin

Volume 2: LINEAR OPERATORS AND OPERATOR EQUATIONS—1971

The Existence of Solutions of Linear Hamiltonian Equations with Unbounded Operators • V. I. Derguzov and V. A. Yakubovich

Resonance in the Oscillation of Linear Systems Under the Action of an Almost Periodic Parametric Perturbation • V. N. Fomin

Some Properties of Transformers Defined by Double-Integral Operators • K. Töllner

On Multidimensional Integral Operators • B. S. Pavlov

On One Class of Stieltjes Multiple-Integral Operators • M. Z. Solomyak and V. V. Sten'kin

On the Propagation of Love Waves Along the Surface of an Inhomogeneous Elastic Body of Arbitrary Shape • V. M. Babich and N. Ya. Kirpichnikova

The Dirichlet Problem for Two-Dimensional Quasi-Linear Second-Order Elliptic Equations • N. M. Ivochkina

On Carleson's Uniqueness Theorem for Analytic Functions with a Finite Dirichlet Integral • V. G. Maz'ya and V. P. Khavin

PROBLEMS IN MATHEMATICAL ANALYSIS · VOLUME 2

Linear Operators and Operator Equations

Edited by
Academician V. I. Smirnov
Leningrad State University

Translated from Russian

CONSULTANTS BUREAU · NEW YORK-LONDON · 1971

The original Russian text was published by Leningrad University Press in 1969 in Leningrad and has been extensively revised and corrected by the editor for this edition. The English translation is published under an agreement with Mezhdunarodnaya Kniga, the Soviet book export agency

Library of Congress Catalog Card Number 68-28092

ISBN 978-1-4757-0015-2 ISBN 978-1-4757-0013-8 (eBook)
DOI 10.1007/978-1-4757-0013-8

A Division of Plenum Publishing Corporation
227 West 17th Street, New York, N.Y. 10011
Softcover reprint of the hardcover 1st edition 1971
United Kingdom edition published by Consultants Bureau, London
A Division of Plenum Publishing Company, Ltd.
Donington House, 30 Norfolk Street, London, W.C. 2, England

CONTENTS

THE EXISTENCE OF SOLUTIONS OF LINEAR HAMILTONIAN EQUATIONS WITH UNBOUNDED OPERATORS

V. I. Derguzov and V. A. Yakubovich

Introduction

The present article is devoted to a proof of the existence of solutions of the linear Hamiltonian equation

$$J\frac{dx}{dt} = H(t)x \quad (H(t+\tau) = H(t)) \tag{1}$$

in a complex separable Hilbert space. Here, J is a symmetric anti-Hermitian operator that is bounded together with its inverse and H(t) is an unbounded operator that, generally speaking, is self-adjoint and close in a definite sense to a positive definite operator.*

Many partial differential equations which describe the oscillation of systems with distributed parameters can be written in the form of an operator differential equation, for example, in the form of the hyperbolic operator equation

$$\frac{d^2v}{dt^2} + P(t)v = 0,$$

in which P(t) is a self-adjoint operator. After a suitable transformation, the latter equation reduces to Eq. (1).

The Cauchy problem of the existence of solutions of operator differential equations has been solved by many authors (see the review articles [1, 2]). Despite the extensive literature on the subject, existence theorems have not been formulated for Eq. (1). The present article contains proofs and a further development of the results obtained by the authors [3] concerning Eq. (1) and differential operator equations of the second order in t.

In most articles, differential equations in abstract spaces are solved by means of a construction of the resolving operator in the form of a multiplicative integral. Of the articles in which this approach has been adopted, that by T. Kato [4] stands out because of the generality of its results. In the present article, the solution of Eq. (1) is also constructed in the form of

*The restrictions imposed on H(t) are rigorously formulated below.

a multiplicative integral. This becomes possible after the construction of an explicit solution for the equation with constant coefficients. Although the solution is given by a multiplicative integral, the desired results cannot be obtained by a direct application of the theorems of [4] to Eq. (1) or the equation which is obtained from Eq. (1) by a simple transformation.

O. A. Ladyzhenskaya [5] and M. I. Vishik [6] have developed another method for proving theorems concerning the existence and uniqueness of solutions. In these articles, equations of the second order in t have been considered among others. The results obtained in the present article for operator differential equations of the second order in t are close to those of [5, 6]. By contrast with [6], the restrictions on the coefficients of the equations are here formulated in explicit form.

§ 1. Preliminary Considerations

Let W be a complex separable Hilbert space. The scalar product of two elements x and y belonging to W is denoted by (x, y) and the norm by $\|x\| = (x, x)^{1/2}$.

Let us consider on a finite segment $[a \le t \le b]$ the collection W_a^b of all measurable functions x(t) with values in W having a finite integral $\int_a^b \|x(t)\|^2 dt$. In this linear set, let us define the scalar product

$$((x(t),\ y(t))) = \int_a^b (x(t),\ y(t))\,dt$$

and the norm

$$|x(t)|_a^b = \left(\int_a^b \|x(t)\|^2 dt\right)^{1/2}.$$

As is well known (see [7], Chapter III), W_a^b will then become a complete separable Hilbert space.

The weak convergence of a sequence of elements x_n to x in space $W\,(W_a^b)$ will be denoted by

$$x_n \xrightarrow{W} x \quad \left(x_n \xrightarrow{W_a^b} x\right).$$

We will say that a weakly absolutely continuous function x(t) has a derivative in W_a^b if the function x(t) is strongly differentiable for almost all t and $\frac{dx(t)}{dt} \in W_a^b$. It is well known that for a weakly absolutely continuous function x(t) to have a derivative in W_a^b, it is necessary and sufficient that it be representable by a Bochner integral

$$x(t) = x(a) + \int_a^t y(s)\,ds,$$

where $y \in W_a^b$. Moreover, $y(t) = \frac{dx(t)}{dt}$ for almost all t. It is easy to see that if functions x(t) and y(t) are weakly absolutely continuous and have derivatives in W_a^b, then we have

$$\int_a^b \left(\frac{dx}{dt}, y\right) dt = -\int_a^b \left(x, \frac{dy}{dt}\right) dt + (x(t), y(t))\Big|_{t=a}^{t=b}.$$

For each fixed t, let A(t) be a bounded operator in W. We will say that it is strongly or weakly continuous at the point t_0 if the function $A(t)x$, with $x \in W$, is strongly or weakly continuous at the point t_0. The operator A(t) is continuous on an interval if it is continuous at every point of this interval.

A bounded operator defined for almost all t is said to be measurable on $[a, b]$ if the function $A(t)x$, with $x \in W$ is measurable. If the operator A(t) and function x(t) are measurable, then the function A(t)x(t) is also measurable. By the integral $\int_a^b A(t)\,dt$ or a measurable operator A(t) we mean the operator A generated by the formula $Ax = \int_a^b A(t)x\,dt$. The integral on the right-hand side of this formula, if it exists, is taken in the Bochner sense.

§ 2. Hamiltonian Equations with Constant Operator Coefficients

Some of the simplest properties of the solutions of a Hamiltonian equation with constant coefficients are obtained in this section. These properties will be used in the following sections for proving the existence of solutions of Hamiltonian equations with variable coefficients.

We will consider in space W the Hamiltonian equation

$$J\frac{dx(t)}{dt} = Hx(t) \tag{2.1}$$

with the initial condition

$$x(t)\big|_{t=a} = x_0. \tag{2.2}$$

Here, J is a linear anti-Hermitian operator ($J^* = -J$) that, together with its inverse, is bounded, and H is a linear self-adjoint positive definite and, in general, unbounded operator; $H^* = H \geq \alpha I$, where $\alpha > 0$ is constant and I is the unit operator.

The Cauchy problem (2.1) and (2.2) is solved on any finite interval $a \leq t \leq b$.

If the initial value x_0 belongs to $D(H^{1/2})$, the domain of definition of the positive square root of the operator H, then we can prove the existence of a generalized solution of problem (2.1) and (2.2).

THEOREM 1. *Let all of the conditions mentioned above concerning the operators J and H be satisfied. If $x_0 \in D(H^{1/2})$, then there exists a unique measurable function x(t) which possesses the following properties:*

1. The function $H^{1/2}x(t)$ has meaning and is bounded uniformly with respect to $t \in [a, b]$.

2. For any element $y \in D(H^{1/2})$ the function x(t) satisfies the integral identity

$$(Jx(t), y) - (Jx(a), y) = \int_a^t (H^{1/2}x(s), H^{1/2}y)\,ds \tag{2.3}$$

for all $t \in [a, b]$.

The function x(t) possessing the above properties is of the form

$$x(t) = H^{-1/2} e^{iK(t-a)} H^{1/2} x_0, \tag{2.4}$$

where

$$K = -iH^{1/2} J^{-1} H^{1/2} \tag{2.5}$$

is a self-adjoint operator.

We will call the function x(t) with properties 1 and 2 of Theorem 1 the generalized solution of problem (2.1-2.2).

Let us prove Theorem 1.

The generally unbounded operator (2.5) is self-adjoint. This follows from the symmetry and boundedness of the inverse operator $K^{-1} = iH^{-1/2}JH^{-1/2}$. The operator $e^{iK(t-a)}$ considered as a function of the self-adjoint operator K is unitary. Therefore, under the conditions of Theorem 1, the function (2.4) is defined for all t.

Let us check that the function $H^{-1/2}Jx(t)$ is strongly differentiable and that the relation

$$\frac{d}{dt}\left(H^{-1/2} J x(t)\right) = H^{1/2} x(t) \tag{2.6}$$

is satisfied. On the basis of (2.4) and (2.5), we have

$$H^{-1/2} J x(t) = -iK^{-1} e^{iK(t-a)} H^{1/2} x_0.$$

Consequently, we must check that the operator $-iK^{-1}e^{iK(t-a)}$ is strongly differentiable or, what is the same, prove the existence of a strong limit for the operators

$$-iK^{-1}\frac{1}{\Delta t}\left[e^{iK(t+\Delta t-a)} - e^{iK(t-a)}\right] \tag{2.7}$$

as $\Delta t \to 0$.

Let us take $E^{(\lambda)} = E_\lambda - E_{-\lambda}$ (with $\lambda > 0$), where E_λ is the resolution of unity of operator (2.5). On all elements of the form $E^{(\lambda)}y$, with $y \in W$ the operators (2.7) strongly converge to the bounded operator $e^{iK(t-a)}$.

It follows from the obvious relation

$$-iK^{-1}\frac{1}{\Delta t}\left[e^{iK(t+\Delta t-a)} - e^{iK(t-a)}\right] = \frac{1}{\Delta t}\int_t^{t+\Delta t} e^{iK(s-a)}\,ds,$$

in which the integral is to be understood in the strong sense, that the norms of the operators (2.7) are bounded uniformly with respect to Δt. Their strong convergence to the operator $e^{iK(t-a)}$ on the whole of space follows from their convergence on a dense set and the uniform boundedness of the operators (2.7) on the basis of the Banach-Steinhaus theorem (see [8], Chapter VII). We have thereby proved the strong differentiability of the operator $-iK^{-1}e^{iK(t-a)}$ and the realization of the relation

$$\frac{d}{dt}\left[-iK^{-1}e^{iK(t-a)}\right] = e^{iK(t-a)},$$

which is equivalent to the strong differentiability of the function $H^{-1/2}Jx(t)$ and the realization of equality (2.6).

Let y belong to $D(H^{1/2})$. Let us multiply equality (2.6) scalarly on the right by the element $H^{1/2}y$ and integrate the resulting expression over the time

$$\int_a^t \left(\frac{d}{ds} H^{-1/2} Jx(s),\ H^{1/2}y\right) ds = \int_a^t (H^{1/2}x(s),\ H^{1/2}y)\, ds.$$

An integration by parts of the left-hand side of the last equality leads to the identity (2.3).

Property 2 of function (2.4) has been proved. Property 1 is obvious. It remains for us to check that the generalized solution (2.4) is unique. This is proved in Section 4 for a more general case. Theorem 1 has been proved.

§ 3. The Existence of a Generalized Solution of the Hamiltonian Equation with Variable Coefficients

Let us consider the problem of the determination of the solutions of the Hamiltonian equation

$$J\frac{dx(t)}{dt} = H(t)\,x(t), \tag{3.1}$$

satisfying the initial condition

$$x(t)\,|_{t=0} = x_0. \tag{3.2}$$

We assume that J and H(t) are linear operators in W that are subject to some general restrictions.

Here and in the following we assume that J is an anti-Hermitian operator (i.e., $J^* = -J$) that is bounded together with its inverse.

The operator H(t) defined on the interval $t \in [a, b)$ can be written as the sum of two operators

$$H(t) = H_0(t) + H_1(t), \tag{3.3}$$

which are subject to the following conditions:

I. The interval $[a, b)$ can be subdivided by means of the points $a = t_0 < t_1 < t_2 < \ldots < t_n = b$ into a finite number of half-open intervals $[t_{k-1}, t_k)$, where $k = 1, 2, \ldots, n$. On all of these half-open intervals, the operator $H_0(t)$ is self-adjoint and sign definite

$$|H_0(t)| \geqslant \alpha I \quad (\alpha = \text{const} > 0).$$

Here, we have $|H_0(t)| = H_0(t)$ when $H_0(t) > 0$ and $|H_0(t)| = -H_0(t)$ when $H_0(t) < 0$.

II. $D(|H_0|^{1/2})$, the domain of definition of the positive square root $|H_0(t)|^{1/2}$ of the positive definite operator $|H(t)|$, is independent of t.

III. The bound

$$\| |H_0(t)|^{1/2} |H_0(s)|^{-1/2} \| \leqslant 1 + \text{const}\,|t-s| \tag{3.4}$$

is satisfied for all t, s belonging to $[t_{k-1}, t_k)$, where $k = 1, 2, \ldots, n$.

IV. For almost all t, the symmetric operator $H_1(t)$ is defined on $D(|H_0|^{1/2})$. The bounded operator

$$A[H_1(t)] = \frac{1}{i} | H_0(a) |^{1/2} J^{-1} H_1(t) | H_0(a) |^{-1/2}, \tag{3.5}$$

which is measurable, has meaning and its norm is Lebesgue integrable:

$$\| A[H_1(t)] \| \in L(a, b).$$

Let us introduce S. L. Sobolev's concept of a generalized solution of problem (3.1) and (3.2).

Definition 1. We will use the term **generalized solution** of problem (3.1)-(3.2) to denote a measurable function x(t) which possesses the following properties:

1. The function $| H_0(a) |^{1/2} x(t)$ has meaning and is bounded uniformly with respect to $t \in [a, b)$.

2. For any $t \in [a, b)$ the integral identity*

$$(Jx(t), y) - (Jx(a), y) = \int_a^t [(|H_0(s)|^{1/2} x(s), |H_0(s)|^{1/2} y) \operatorname{sign} H_0(s) + (H_1(s) x(s), y)] \, ds \tag{3.6}$$

is satisfied for any element y belonging to $D(| H_0 |^{1/2})$.

With the above assumptions problem (3.1)-(3.2) has a generalized solution.

THEOREM 2. If Conditions I-IV are satisfied and $x(a) \in D(|H_0|^{1/2})$, then Eq. (3.1) with the initial condition (3.2) has a unique generalized solution.

Before proceeding to prove Theorem 2, we will show that Condition III is equivalent to any one of the following:

V. For any $t, s \in [t_{k-1}, t_k)$, where k = 1, 2, ..., n, we have

$$\| [| H_0(t) |^{1/2} | H_0(s) |^{-1/2}]^* | H_0(t) |^{1/2} | H_0(s) |^{-1/2} I \| \leqslant \text{const} | t - s |. \tag{3.7}$$

VI. For $t, s \in [t_{k-1}, t_k)$ with k = 1, 2, ..., n, and arbitrary elements x, y belonging to $D(| H_0 |^{1/2})$, we have

$$|(|H_0(t)|^{1/2} x, |H_0(t)|^{1/2} y) - (|H_0(s)|^{1/2} x, |H_0(s)|^{1/2} y)| \leqslant \text{const} \| |H_0(s)|^{1/2} x \| \cdot \| |H_0(s)|^{1/2} y \| \cdot | t - s |. \tag{3.8}$$

Let us show that Condition III implies V, Condition V implies VI, and Condition VI implies III. This will prove that Conditions III, V, and VI are equivalent.

Let us take

$$M(t, s) = | H_0(t) |^{1/2} | H_0(s) |^{-1/2} \tag{3.9}$$

and

$$A(t, s) = M^*(t, s) M(t, s). \tag{3.10}$$

The bounded symmetric operator A(t, s) has a bounded inverse

$$A^{-1}(t, s) = M(s, t) M^*(s, t). \tag{3.11}$$

If Condition III is satisfied, then we have $\| M(s, t) \| \leq 1 + \text{const} |t - s|$. Therefore, from (3.10) and (3.11), we obtain for the positive operator A(t, s) the bound

---

* The numerical function sign in identity (3.6) is defined as follows: sign $H_0(t) = 1$ when $H_0(t) > 0$ and sign $H_0(t) = -1$ when $H_0(t) < 0$.

$$\frac{1}{(1+\text{const}\,|t-s|)^2} I \leqslant A(t, s) \leqslant (1+\text{const}\,|t-s|)^2 I,$$

which yields the inequality

$$\|A(t, s)-I\| \leqslant \text{const}\,|t-s|. \tag{3.12}$$

Inequality (3.12) is identical with (3.7). Therefore, III follows from V.

Applying the operator $A(t, s) - I$ to the element $|H_0(s)|^{1/2}x$ and multiplying the resulting expression scalarly by the element $|H_0(s)|^{1/2}y$, we obtain (3.8) with the help of (3.12). Thus, VI follows from V.

Setting $x = y = |H_0(s)|^{-1/2}z$ in (3.8), we find that

$$\left| \| |H_0(t)|^{1/2} |H_0(s)|^{-1/2} z\|^2 - \|z\|^2 \right| \leqslant (1+\text{const}\,|t-s|)^{1/2}.$$

This leads to the bound

$$\| |H_0(t)|^{1/2} |H_0(s)|^{1/2} \| \leqslant (1+\text{const}\,|t-s|)^{1/2},$$

which is equivalent to (3.4). Thus, III follows from VI.

Let us now proceed to prove Theorem 2. In the present section, we will only give an incomplete proof of Theorem 2 and introduce a number of additional restrictions which will be removed in the following.

Let us make the following additional assumptions: $H_1(t) \equiv 0$, the first half-open interval $[a, t_1)$ coincides with the whole of the half-open interval $[a, b)$, and $H_0(t)$ is positive definite on $[a, b)$. The existence of a generalized solution will now be proved in the presence of the above additional restriction.

We divide the segment $[a, b]$ into n equal parts by the points

$$t_k = a + \frac{b-a}{n} k \quad (k = 0, 1, \dots, n).$$

In each segment $[t_k, t_{k+1}]$ we replace the operator $H(t)$ in Eq. (3.1) by the operator $H(t_k)$ and solve the resulting equations segment by segment. We take the initial value on the first segment $[a, t_1]$ to be $x_0 \in D(|H_0|^{1/2})$. On each following segment $[t_k, t_{k+1}]$, the initial value is given by the solution of the equation at the point t_k on the preceding segment. Following this procedure, we find on the basis of Theorem 1 that the equation

$$J\frac{dx}{dt} = H(t_k)x \qquad (t_k \leqslant t \leqslant t_{k+1}) \tag{3.13}$$

is solved in the generalized sense by the function

$$x_n^{(k)}(t) = H^{-1/2}(t_k) e^{iK(t_k)(t-t_k)} H^{1/2}(t_k) x_n^{(k-1)}(t_k) \tag{3.14}$$

with the initial value

$$x_n^{(k-1)}(t_k) = H^{-1/2}(t_{k-1}) e^{iK(t_{k-1})(t_{k-1}-t_k)} H^{1/2}(t_{k-1}) \times \dots \times H^{-1/2}(a) e^{iK(a)(t_1-a)} H^{1/2}(a) x(a), \tag{3.15}$$

belonging to $D(|H_0|^{1/2})$, provided that $x_0 \in D(|H_0|^{1/2})$.

In the above formulas we have $K(t) = -iH_0^{1/2}(t) J^{-1} H_0^{1/2}(t)$. Also on the basis of Theorem 1 of Section 2, the function (3.14) satisfies on the segment $t \in [t_k, t_{k+1}]$ the integral identity

$$(Jx_n^{(k)}(t), y) - (Jx_a^{(k)}(t_k), y) = \int_{t_k}^{t} (H_0^{1/2}(t_k) x_n^{(k)}(s), H_0^{1/2}(t_k) y)\, ds \tag{3.16}$$

for any element $y \in D(|H_0|^{1/2})$.

Let the function $x_n(t)$ and operator $H_n(t)$ be equal to $x_n^{(k)}(t)$ and $H_0(t_k)$, respectively, for $t\in[t_k, t_{k+1}]$, where $k = 0, 1,\dots, n-1$.

Let t be any number in $[a, b)$. This t belongs to some half-open interval $[t_{k_0}, t_{k_0+1})$. Let us set $t = t_{k+1}$ for all $k < k_0$ in expressions (3.16) and perform a term-by-term addition of them and expression (3.16) with $k = k_0$; we will then obtain

$$(Jx_n(t), y)-(Jx(a), y)=\int_a^t (H_n^{1/2}(s)x_n(s), H_n^{1/2}(s)y)\,ds. \tag{3.17}$$

Hence, for any $t\in[a, b)$ the function $x_n(t)$ satisfies the integral identity (3.17) for an arbitrary $y\in D(|H_0|^{1/2})$.

Using (3.4), we can find from formulas (3.14) and (3.15) that

$$\|H_0^{1/2}(a)x_n(t)\|\leqslant\|H_0^{1/2}(a)H_n^{-1/2}(t)\|\prod_{k=0}^{n-1}(1+\text{const}\,|t_{k+1}-t_k|)\cdot\|H_0^{1/2}(a)x(a)\|. \tag{3.18}$$

The quantity $\|H_0^{1/2}(a)H_n^{-1/2}(t)\|$ is bounded in view of (3.4) and the product $\prod_{k=1}^{n}(1+\text{const}\,|t_{k+1}-t_k|)$ $=\prod_{k=1}^{n}\left(1+\frac{\text{const}\,|b-a|}{n}\right)_k$ has a finite limit equal to $\exp[\text{const}\,|b-a|]$. It follows from this that the function $H_0^{1/2}(a)x_n(t)$ is uniformly bounded,

$$\|H_0^{1/2}(a)x_n(t)\|\leqslant\text{const}\,\|H_0^{1/2}(a)x(a)\|, \tag{3.19}$$

and even that functions $x_n(t)$ are uniformly bounded.

Hilbert space W_a^b is weakly compact. Consequently, on a subsequence of numbers (this subsequence will again be denoted by $\{n\}$), the bounded functions $x_n(t)$ and $H_0^{1/2}(a)x_n(t)$ weakly converge in W_a^b to the functions $x(t)$ and $H_0^{1/2}(a)x(t)$, an event written as

$$x_n(t)\xrightarrow{W_a^b}x(t),\quad H_0^{1/2}(a)x_n(t)\xrightarrow{W_a^b}H_0^{1/2}(a)x(t). \tag{3.20}$$

The following equalities are obvious:

$$\begin{gathered}H_n^{1/2}(t)x_n(t),\ H_n^{1/2}(t)y)=(H_0^{1/2}(a)x_n(t),\ [H_n^{1/2}(t)H_0^{-1/2}(a)]^*H_n^{1/2}(t)H_0^{-1/2}(a)y),\\ [H_n^{1/2}(t)H_0^{-1/2}(a)]^*H_n^{1/2}(t)H_0^{-1/2}(a)-[H_0^{1/2}(t)H_0^{-1/2}(a)]^*H_0^{1/2}(t)H_0^{-1/2}(a)=\\ =[H_0^{1/2}(t)H_0^{-1/2}(a)]^*\{[H^{1/2}(t)H_0^{-1/2}(t)]^*H_n^{1/2}(t)H_0^{-1/2}(t)-I\}H_0^{1/2}(t)H_0^{-1/2}(a).\end{gathered} \tag{3.21}$$

In view of (3.7), the last equality yields

$$[H_n^{1/2}(t)H_0^{-1/2}(a)]^*H_n^{1/2}(t)H_0^{-1/2}(a)y-[H_0^{1/2}(t)H_0^{-1/2}(a)]^*H_0^{1/2}(t)H_0^{-1/2}a(y)\|\xrightarrow[n\to\infty]{}0 \tag{3.22}$$

uniformly with respect to $t\in[a, b)$.

It is not difficult to show with the help of (3.17), (3.19), and (3.20) that on some subsequence of numbers, which we will again denote by $\{n\}$, the functions $x_n(t)$ weakly converge to the function $x(t)$ for almost all t:

$$x_n(t)\xrightarrow{W}x(t). \tag{3.23}$$

Proceeding to the limit in identity (3.17) and taking (3.21)-(3.23) into account, we find that the function x(t) satisfies the integral identity

$$(Jx(t), y) - (Jx(a), y) = \int_a^t (H_0^{1/2}(s)\,x(s),\ H_0^{1/2}(s)\,y)\,ds, \tag{3.24}$$

valid for almost all $t \in [a, b]$.

In view of (3.19), the limiting function x(t) satisfies the inequality

$$\| H_0^{1/2}(a)\,x(t)\| \leqslant \text{const}\,\| H_0^{1/2}(a)\,x(a)\| \tag{3.25}$$

for almost all t.

The right-hand side of identity (3.24) depends continuously on t and completes the definition of the function (Jx(t), y) with respect to continuity. Since x(t) is uniformly bounded with respect to t, as follows from (3.25), and since the set of elements of $\{Jy\}$ is dense in W when $y \in D(|H_0|^{1/2})$, the functional (Jx(t), y) completes for all t the definition of x(t) with respect to weak continuity, provided that x(t) is replaced by an equivalent weakly continuous function.

The function $H_0^{1/2}(a)x(t)$ has meaning for almost all t and on account of (3.25) it is uniformly bounded with respect to t. It follows from the weak continuity of x(t) and the uniform boundedness of $H_0^{1/2}(a)x(t)$ for almost all t that $x(t) \in D(|H_0|^{1/2})$ for all $t \in [a, b]$, and that the function $H_0^{1/2}(a)x(t)$ is weakly continuous. It is now clear that the bound (3.25) holds for all $t \in [a, b]$.

With the additional assumptions introduced above concerning the operators $H_0(t)$ and $H_1(t)$, we have proved the existence of a function x(t) described in Definition 1, i.e., we have proved the existence of a generalized solution.

The additional assumptions were that $H_0(t) > 0$ on the semi-open interval $[a, b) = [a, t_1)$ and that $H_1(t) \equiv 0$.

Let us now assume that $H_0(t) < 0$ on $[a, b) = [a, t_1)$ and that $H_1(t) \equiv 0$. We can reduce this case to the preceding one if we multiply Eq. (3.1) by -1 and replace the operator J by $-J$. The existence of a generalized solution of Eq. (3.1) in the sense of Definition 1 follows directly from this.

If we now assume that $H_1(t) \equiv 0$ and that $H_0(t)$ satisfies the condition of Theorem 2, then the above shows that a generalized solution exists on every semi-open interval $[t_k, t_{k+1})$. Let us construct on $[a, b)$ a function x(t) equal on the semi-open interval $[t_k, t_{k+1})$ to the generalized solution of Eq. (3.1) with the initial condition $x(t_k)$ which is equal to the value at the point $t = t_k$ of the generalized solution of Eq. (3.1) on the preceding semi-interval $[t_{k-1}, t_k)$. The function x(t) constructed in this way will satisfy the integral identity (3.6) because x(t) is a generalized solution on any semi-open interval $[t_k, t_{k+1})$. Thus, x(t) is a generalized solution.

In the following section we will prove the existence and uniqueness of the generalized solution also when $H_1(t) \not\equiv 0$, i.e., we will complete the proof of Theorem 2.

§ 4. The Existence and Uniqueness of the Generalized Solution of the Hamiltonian Equation and the Simplest Properties of the Solution

Let us prove the uniqueness of the generalized solution of problem (3.1)-(3.2) on the assumption that this solution exists. Let us set $y = |H_0(a)|^{-1/2}z$ in identity (3.6), so that we have

$$(|H_0(a)|^{-1/2} Jx(t),\ z) - (|H_0(a)|^{-1/2} Jx(a),\ z) =$$

$$= \int_a^t \{\operatorname{sign} H_0(s) \cdot ([|H_0(s)|^{1/2} |H_0(a)|^{-1/2}]^* |H_0(s)|^{1/2} x(s),\ z) + (|H_0(a)|^{-1/2} H_1(s) x(s),\ z)\}\, ds,$$

where $[|H_0(s)|^{1/2} |H_0(a)|^{1/2}]^*$ is the operator that is the conjugate of the operator $|H_0(s)|^{1/2} |H_0(a)|^{-1/2}$.

The last identity can be interpreted as an integral equation,

$$|H_0(a)|^{-1/2} Jx(t) = |H_0(a)|^{-1/2} Jx(a) + \\ + \int_a^t \{(\operatorname{sign} H_0(s)) \cdot [|H_0(s)|^{1/2} |H_0(a)|^{-1/2}]^* |H_0(s)|^{1/2} x(s) + |H_0(a)|^{-1/2} H_1(s) x(s)\}\, ds, \tag{4.1}$$

in which, following Pettis [7], we must take the integral in its weak sense. Since the integrand is summable, the integral in (4.1) can be taken in the strong sense according to Bochner. Therefore, the function $H_0(a)^{-1/2}Jx(t)$ is strongly differentiable for almost all t and satisfies the equation

$$\frac{d}{dt}[|H_0(a)|^{-1/2} Jx(t)] = |H_0(a)|^{-1/2} H_1(t) x(t) + (\operatorname{sign} H_0(t) \cdot [|H_0(t)|^{1/2} (H_0(a)|^{-1/2}]^* |H_0(t)|^{1/2} x(t). \tag{4.2}$$

Let x(t) and y(t) be two generalized solutions of Eq. (3.1) with distinct initial values belonging to $D(|H_0(a)|^{1/2})$. Using the properties of the generalized solution and Eq. (4.1), we can easily verify that the numerical function (Jx(t), y(t)) is absolutely continuous. Let us calculate the derivative of this function and for this purpose let us write down the difference equation

$$\frac{1}{\Delta t}[(Jx(t+\Delta t),\ y(t+\Delta t)) - (Jx(t),\ y(t))] = \frac{1}{\Delta t}[(J[x(t+\Delta t) - x(t)],\ y(t+\Delta t)) + (Jx(t),\ [y(t+\Delta t) - y(t)])] =$$

$$= \frac{1}{\Delta t}[(|H_0(a)|^{-1/2} J[x(t+\Delta t) - x(t)],\ |H_0(a)|^{1/2} y(t+\Delta t)) - (|H_0(a)|^{1/2} x(t),\ |H_0(a)|^{-1/2} J[y(t+\Delta t) - y(t)])].$$

The function $|H_0(a)|^{1/2}y(t)$ is weakly continuous as follows from the fact that this function is uniformly bounded and that the function $|H_0(a)|^{-1/2}Jy(t)$ is continuous on account of equality (4.1). The function $|H_0(a)|^{-1/2}Jx(t)$ is strongly differentiable for almost all t and satisfies Eq. (4.2). Therefore, the derivative of the function (Jx(t), y(t)) is zero for almost all t:

$$\frac{d}{dt}(Jx(t),\ y(t)) = 0.$$

This means that any two generalized solutions x(t) and y(t) satisfy the relation

$$(Jx(t),\ y(t)) = \text{const}. \tag{4.3}$$

Let us verify the uniqueness of the generalized solution of problem (3.1)-(3.2) with the help of identity (4.3). Let x(t) be the generalized solution corresponding to the initial condition $x(a) = 0$. We must show that $x(t) \equiv 0$. Let us assume that $t_0 \in (a, b)$. We take an arbitrary element $y(t_0)$ belonging to $D(|H_0(a)|^{1/2})$. With the initial value $y(t_0)$ at the point t_0, we solve Eq. (3.1) in the generalized sense on the intervals $[a, t_0]$ and $[t_0, b]$. Let us denote these solutions by $y_1(t)$ (where $t \in [a, t_0]$) and $y_2(t)$ (where $t \in [t_0, b]$). The function

$$y(t) = \begin{cases} y_1(t), & t \in [a, t_0], \\ y_2(t), & t \in [t_0, b], \end{cases}$$

is obviously a generalized solution of Eq. (3.1) on $[a, b]$ that assumes the given value $y(t_0)$ at the point t_0. Making use of (4.3) we have

$$(Jx(t_0),\ y(t_0)) = (Ix(a),\ y(a)) = 0 \qquad (x(a) = 0).$$

Since the element $y(t_0)$ is arbitrary, we see from this that $x(t_0) = 0$. Since t_0 is an arbitrary point of (a, b), we have $x(t) \equiv 0$.

On the assumption of the existence of a generalized solution, we have proved that it is unique.

The existence of a generalized solution of Eq. (3.1) for $H_1(t) \equiv 0$ has been proved in Section 3. Consequently, a theorem of the existence and uniqueness of the generalized solution of Eq. (3.1) holds in this case.

We will temporarily assume that $H_1(t) \equiv 0$ in Eq. (3.1). Let us use the generalized solution $x(t)$ to define the resolving operator $X_0(t)$ as follows:

$$X_0(t)\,x(a) = x(t).$$

On account of the properties listed above, the operator $X_0(t)$ maps the set $D(|H_0|^{1/2})$ onto itself. Because of the bound (3.25), the operator

$$Z_0(t) = |H_0(a)|^{1/2} X_0(t) |H_0(a)|^{-1/2}$$

is uniformly bounded with respect to t. The inverse operator $X_0^{-1}(t)$ exists and possesses all of the abstract properties of $X_0(t)$, i.e., $X_0^{-1}(t)$ maps the set $D(|H_0|^{1/2})$ onto itself, so that $Z_0^{-1}(t)$ is uniformly bounded with respect to $t \in [a, b)$.

When $H_1(t) \not\equiv 0$ we seek the resolving operator $X(t)$ of Eq. (3.1) in the form

$$X(t) = |H_0(a)|^{-1/2} Z(t) |H_0(a)|^{1/2}. \tag{4.4}$$

In order to find $X(t)$, we solve the integral equation

$$X(t) = X_0(t)\left[I + \int_a^t X_0^{-1}(s) H_1(s) X(s)\, ds\right] \tag{4.5}$$

or, in terms of the operator $Z(t)$, the equation

$$Z(t) = Z_0(t)\left[I + i\int_a^t Z_0^{-1}(s) \cdot A[H_1(s)] \cdot Z(s)\, ds\right], \tag{4.6}$$

where the operator $A[H_1(t)]$ is defined by formula (3.5). Equation (4.6) has a unique solution which is uniformly bounded with respect to t,

$$\| Z(t) \| \leqslant \| Z_0(t) \| \exp\left[\int_a^t \| Z_0^{-1}(s) \| \cdot \| A[H_1(s)] \| \cdot \| Z_0(s) \| ds\right].$$

The inverse operator $Z^{-1}(t)$ exists and is obviously also uniformly bounded with respect to $t \in [a, b)$.

Using the given initial condition $x(a) \in D(|H_0|^{1/2})$ and the operator (4.4) constructed above, we define the function

$$X(t) = X(t)\,x(a). \tag{4.7}$$

It can be easily shown that function (4.7) is the generalized solution of problem (3.1)-(3.2). The existence of a generalized solution of problem (3.1)-(3.2) under the conditions of Theorem 2 has been proved. The uniqueness of such a solution has been proved at the beginning of the present section. Thus, Theorem 2 has been completely proved.

Along the way, we have proved the properties ofthe resolving operator and the generalized solution that are listed below in Theorem 3 and Definition 2.

THEOREM 3. If the conditions of Theorem 2 are satisfied, then the resolving operator $X(t)$ of Eq. (3.1) and its inverse $X^{-1}(t)$ map the set $D(|H_0|^{1/2})$ onto itself. The operator $Z(t) = |H_0(a)|^{1/2}X(t)|H_0(a)|^{-1/2}$ and its inverse are uniformly bounded with respect to $t \in [a, b)$.

Definition 2. We will use the term generalized solution of Eq. (3.1) to denote any measurable function $x(t)$ that possesses the following properties:

1. The function $|H_0(a)|^{1/2}x(t)$ has meaning and is weakly continuous with respect to t.

2. The strong derivative $\frac{d}{dt}|H_0(a)|^{-1/2}Jx(t)$ exists for almost all t and for these t the function $x(t)$ satisfies Eq. (4.2).*

As has been shown, the existence of a generalized solution in the sense of Definition 1 implies its existence in the sense of Definition 2. The converse can be shown by a scalar multiplication of Eq. (4.2) by a constant vector and an integration by parts. Consequently, we have

THEOREM 4. If the conditions of Theorem 2 are satisfied, then the existence and uniqueness of the generalized solution in the sense of either Definition 1 or Definition 2 implies its existence in the sense of the other definition.

Let us introduce a bounded symmetric operator

$$F = -i|H_0(a)|^{-1/2}J|H_0(a)|^{-1/2}, \tag{4.8}$$

which, in general, has an unbounded inverse. In view of (4.3), for any elements x, y belonging to $D(|H_0|^{1/2})$ we have

$$(JX(t)x,\ X(t)y) = (Jx,\ y)$$

or, because of (4.4),

$$(J|H_0(a)|^{-1/2}Z(t)|H_0(a)|^{1/2}x,\ |H_0(a)|^{-1/2}Z(t)|H_0(a)|^{1/2}y) = (Jx,\ y).$$

Adopting the abbreviations $x' = |H_0(a)|^{1/2}x$ and $y' = |H_0(a)|^{1/2}y$, we find from this relation that

$$(FZ(t)x',\ Z(t)y') = (Fx',\ y'),$$

or

$$Z^*(t)FZ(t) = F. \tag{4.9}$$

Thus, we have proved

THEOREM 5. If the conditions of Theorem 2 are satisfied, then the operator $Z(t)$ defined in Theorem 3 satisfies (4.9) in which the operator F is defined by formula (4.8).

---

*The function $|H_0(a)|^{-1/2}Jx(t)$ is assumed to be weakly absolutely continuous.

Finally, let us show that the operator Z(t) depends continuously on $H_1(t)$ for small changes in the operator $H_1(t)$ in a special topology.

Let the symmetric operator $\tilde{H}_1(t)$ satisfy Condition IV of Section 3, i.e., a bounded measurable operator

$$A[\tilde{H}_1(t)] = \frac{1}{i} | H_0(a) |^{1/2} J^{-1} \tilde{H}_1(t) | H_0(a) |^{-1/2}, \tag{4.10}$$

whose norm is Lebesgue summable on the segment $[a, b]$, is defined for almost all t.

From what has been proved above, we see that the resolving operator of the equation

$$J \frac{dx}{dt} = [H_0(t) + \tilde{H}_1(t)] x \tag{4.11}$$

is of the form $\tilde{X}(t) = | H_0(a) |^{-1/2} \tilde{Z}(t) | H_0(a) |^{1/2}$. It is obvious that the bounded operators Z(t) and $\tilde{Z}(t)$ are related by the integral equation

$$\tilde{Z}(t) = Z(t) \left[I + i \int_a^t Z^{-1}(s) \cdot A[\tilde{H}_1(s) - H_1(s)] \right] \cdot \tilde{Z}(s)\, ds. \tag{4.12}$$

This leads to the bound

$$\| \tilde{Z}(t) - Z(t) \| \leqslant \| Z(t) \| \left[\exp \left(\int_a^t \| Z^{-1}(s) \| \cdot \| A[\tilde{H}_1(s) - H_1(s)] \| \| Z(s) \| ds \right) - 1 \right]. \tag{4.13}$$

Let the operator $H_0(t)$ be fixed and let **H** be the set of all operators $H_1(t)$ satisfying Condition IV of Section 3. Let us introduce in **H** the distance between operators $H_1 = H_1(t)$ and $\tilde{H}_1 = \tilde{H}_1(t)$ by means of the formula

$$\rho(H_1, \tilde{H}_1) = \int_a^b \| A[\tilde{H}_1(t)] - A[H_1(t)] \| dt, \tag{4.14}$$

where the operators $\tilde{A}[H_1(t)]$ and $A[H_1(t)]$ are defined by formulas (4.10) and (3.5), respectively.

When the fact that the operators Z(t) and $Z^{-1}(t)$ are uniformly bounded with respect to t is taken into account, inequality (4.13) yields.

THEOREM 6. *Let the coefficients of Eq. (3.1) satisfy the conditions of Theorem 2 and let* $X(t) = | H_0(a) |^{-1/2} Z(t) | H_0(a) |^{1/2}$ *be the resolving operator. Then, small changes in the operator* $H_1(t) \in \mathbf{H}$ *in topology (4.14) correspond to small changes in the operator Z(t) in the uniform operator topology.*

§ 5. On the Smoothness of the Generalized Solutions of the Hamiltonian Equations

As the smoothness of the coefficients of Eq. (3.1) of the initial condition (3.2) increases, the smoothness properties of the generalized solution improve.

We will assume that in addition to Properties I and II of Section 3, the following conditions are imposed in the coefficients of Eq. (3.1):

VII. The operator $T(t)=|H_0(t)|^{1/2}J^{-1}|H_0(t)|$ has a constant domain of definition and the following bound is valid for arbitrary $t, s\in[t_k, t_{k+1})$, with k = 1, 2, ..., n:

$$\|T(t)T^{-1}(s)\|\leqslant 1+\text{const}|t-s|. \tag{5.1}$$

VIII. The bounded operator

$$\Gamma(t)=|H_0(a)|^{1/2}J^{-1}|H_0(a)|J^{-1}H_1(t)|H_0(a)|^{-1}J|H_0(a)|^{-1/2},$$

measurable with respect to t, has meaning for almost all t and its norm is Lebesgue summable:

$$\|\Gamma(t)\|\in L(a, b). \tag{5.2}$$

When $x(a)\in D(T)$ and the conditions concerning the coefficients of Eq. (3.1) listed above are satisfied, the generalized solution becomes an ordinary solution. This result is formulated in the following theorem:

THEOREM 7. If Conditions I, VII, and VIII are satisfied* and $x(a)\in D(T)$, then the generalized solution of problem (3.1)-(3.2) possesses the following properties:

1. x(t) belongs to D(t) for all $t\in[a, b]$ and we have

$$\|T(a)x(t)\|\leqslant\text{const}\|T(a)x(a)\|. \tag{5.3}$$

2. For almost all t the function x(t) is strongly differentiable and satisfies Eq. (3.1).

Before proving Theorem 7, we will show that Conditions VII and VIII imply Conditions II, III, and IV if Condition I is satisfied.

We clearly have

$$T(t)T^{-1}(s)=A(t, s)F^{-1}(s)A^*(t, s)A(t, s)F(s), \tag{5.4}$$

where the operator A(t, s) is defined on the dense set

$$A(t, s)=|H_0(t)|^{1/2}|H_0(s)|^{-1/2},$$

and F(s) is a bounded symmetric operator

$$F(s)=-i|H_0(s)|^{-1/2}J|H_0(s)|^{-1/2}.$$

Using the obvious properties of the operator $A(s, t) = A^{-1}(t, s)$, we find from (5.4) that

$$\|F^{-1}(s)A^*(t, s)F(s)\|\leqslant\|T(t)T^{-1}(s)\|\cdot\|A(s, t)\|. \tag{5.5}$$

According to Lemma 3 of [9], this leads to the bound

$$\|A^*(t, s)A(t, s)\|\leqslant\|T(t)T^{-1}(s)\|\cdot\|A(s, t)\|. \tag{5.6}$$

Combining (5.6) with a bound like (5.6) except that t and s are interchanged, we obtain

$$\|A(t, s)\|^3\leqslant\|T(t)T^{-1}(s)\|^2\|T(s)T^{-1}(t)\|$$

or, with the help of (5.1),

$$\|A(t, s)\|\leqslant 1+\text{const}|t-s|.$$

*Conditions I, VII, VIII imply conditions II, III, IV and, consequently, all conditions of Theorem 2 are satisfied, so that the generalized solution exists.

Hence, the operator A(t, s) is defined on a dense set and is bounded. Therefore, $D(|H_0(t)|^{1/2})$ is independent of t and (3.4) holds.

Operator (5.2) can be represented as

$$\Gamma(t) = iF^{-1}A[H_1(t)]F.$$

The operators $A[H_1(t)]$ and F are defined by formulas (3.5) and (4.8), respectively. The easily checked formal relation $FA[H_1(t)]F^{-1} = -A^*[H_1(t)]$ yields the equality

$$iF^2A[H_1(t)]F^{-2} = \Gamma^*(t), \tag{5.7}$$

which is valid on a dense set $D(F^3)$. Let us introduce the positive definite operator $C = |F|^{-3}$, where $|F| = (F \cdot F^*)^{1/2}$, and the operator $B(t) = |F|^2A[H_1(t)]|F|^{-2}$. We obtain from (5.7) the following bound for the operator B(t) which is defined on the set $D(F^3)$:

$$\|B(t)\| \leqslant \|\Gamma(t)\|. \tag{5.8}$$

We obviously have

$$\|CB(t)C^{-1}\| \leqslant \|\Gamma(t)\|. \tag{5.9}$$

Inequalities (5.8) and (5.9), together with Lemma 4.1 of [1] yield the bound

$$\|C^{\lambda}B(t)C^{-\lambda}\| \leqslant \|\Gamma(t)\|, \tag{5.10}$$

valid for any λ in the interval $0 < \lambda < 1$. When $\lambda = 2/3$, inequality (5.10) yields

$$\|A[H_1(t)]\| \leqslant \|\Gamma(t)\|. \tag{5.11}$$

Thus, the operator $A[H_1(t)]$ has meaning for almost all t and is bounded.

Let us proceed to prove Theorem 7.

Let us initially set $H_1(t) \equiv 0$. In this case, we have determined in Section 3 a sequence of functions $x_n(t)$ which weakly converged to the generalized solution x(t). The functions $x_n(t)$ were defined in terms of functions (3.14) and elements (3.15) which, on the basis of the assumptions made above, can be represented as

$$x_n^{(k)}(t) = T^{-1}(t_k)e^{iK(t_k)(t-t_k)}T(t_k)x_n^{(k-1)}(t_k),$$
$$x_n^{(k-1)}(t_k) = T^{-1}(t_{k-1})e^{iK(t_{k-1})(t_k-t_{k-1})}T(t_{k-1}) \times \ldots \times T^{-1}(a)e^{iK(a)(t_1-a)}T(a)x(a).$$

Using (5.1), we find from the last formulas that

$$\|T(a)x_n(t)\| \leqslant \text{const}\|T(a)x(a)\|,$$

valid for all $t \in [a, b]$.

Since the generalized solution x(t) is unique, the sequence $T(a)x_n(t)$ converges to the function $T(a)x(t)$ for almost all t. In the limit, the last estimate yields (5.3) for almost all t. Since the function x(t) is weakly continuous, it follows from this that $x(t) \in D(T)$ for almost all t and the bound (5.3) holds for all $t \in [a, b]$.

Let us apply the operator $|H_0(a)|^{1/2}$ to relations (4.1); we will then find with the help of (5.3) that the function x(t) is differentiable for almost all t and satisfies Eq. (3.1).

It was assumed above that $H_0(t) > 0$ on $[a, b) = [a, t_1)$. When $H_0(t) < 0$, the proof of Theorem 7 remains in force. In the general case, when $[a, b)$ is subdivided into several half-open intervals $[t_k, t_{k+1})$, the solution is constructed in the same way as in Section 3. Theorem 7 has been proved for $H_1(t) \equiv 0$.

When we have $H_1(t) \not\equiv 0$, the solution x(t) is given by formula (4.7) in which X(t) is the resolving operator. We seek the operator X(t) with the help of the integral equation (4.5) in which $X_0(t)$ is the resolving operator of the same Eq. (3.1) when $H_1(t) \equiv 0$. Inequality (5.3) holds for $H_1(t) \equiv 0$ and it follows that the operator $X_0(t)$ can be represented as

$$X_0(t) = T^{-1}(a) V_0(t) T(a),$$

where $V_0(t)$ is an operator that is uniformly bounded with respect to $t \in [a, b]$ together with its inverse.

We seek X(t) in the same form

$$X(t) = T^{-1}(a) V(t) T(a),$$

so that V(t) is given by the integral equation

$$V(t) = V_0(t) \left[I + \int_a^t V_0^{-1}(s) \Gamma(s) V(s) ds \right].$$

The solution of this equation is an operator that is bounded together with its inverse. It is now easy to verify that the solution x(t) = X(t)x(a) constructed in this manner possesses Properties 1 and 2 of Theorem 7. The theorem has been proved.

§ 6. Second-Order Operator Differential Equations

The results of Sections 2-5 can be extended to an operator equation of the second order in t,

$$\frac{d}{dt}\left[M^{-1}(t) \frac{d}{dt} u(t)\right] + Q \frac{d}{dt} u(t) + P(t) u(t) = 0, \tag{6.1}$$

which is frequently encountered in applied problems. The problem consists in the determination of the solution which at t = a satisfies the initial conditions

$$u(t)|_{t=a} = u(a), \quad \left.\frac{du(t)}{dt}\right|_{t=a} = \dot{u}(a). \tag{6.2}$$

Equation (6.1) will be considered in a complex separable Hilbert space R with the scalar product [x, y] of elements x, y belonging to R and the norm $\|x\|_R = [x, x]^{1/2}$.

In Eq. (6.1), u(t) is a function with values in R, while Q, M(t), and P(t) (for each fixed t) are linear operators in R. P(t) can be written as the sum of two operators

$$P(t) = P_0(t) + P_1(t). \tag{6.3}$$

The following assumptions are made concerning Q, M(t), $P_0(t)$, $P_1(t)$ on the half-open interval [a, b):

IX. Q is a bounded anti-Hermitian operator.

X. The operators M(t) and $P_0(t)$ satisfy Conditions I, II, and III of Section 3 in which $H_0(t)$ is replaced by either M(t) or $P_0(t)$. It is assumed in addition that sign M(t) = sign $P_0(t)$ on every half-open interval $[t_k, t_{k+1})$.

XI. For almost all t, the bounded operator $|M(a)|^{1/2} P_1(t) |P_0(a)|^{-1/2}$, which is measurable, has meaning for a symmetric operator $P_1(t)$ and its norm is Lebesgue integrable on $[a, b]$.

The substitution

$$x(t) = \begin{pmatrix} u(t) \\ M^{-1}(t)\,\dot{u}(t) \end{pmatrix}, \tag{6.4}$$

which for the finite-dimensional case may be found, for example, in [10], can be used to reduce (6.1) and (6.2) to Eq. (3.1) with the initial condition

$$x(a) = \begin{pmatrix} u(a) \\ M^{-1}(a)\,\dot{u}(a) \end{pmatrix}. \tag{6.5}$$

Equation (3.1) is considered in the space $W = R \times R$ whose elements are pairs $x = \{x_1, x_2\}$ and $y = \{y_1, y_2\}$; the scalar product is defined by the formula $(x, y) = [x_1, y_1] + [x_2, y_2]$. The operator coefficients in Eq. (3.1) are

$$J = \begin{pmatrix} -Q & -I \\ I & Q \end{pmatrix}, \quad H_0(t) = \begin{pmatrix} P_0(t) & 0 \\ 0 & M(t) \end{pmatrix},$$
$$H_1(t) = \begin{pmatrix} P_1(t) & 0 \\ 0 & 0 \end{pmatrix}. \tag{6.6}$$

Conditions IX, X, and XI obviously imply Conditions I, II, III, IV. Consequently, if we have $u(a) \in D(|P_0|^{1/2})$ (and this is assumed in the following), then $x(a) \in D(|H_0|^{1/2})$, and problem (3.1)-(3.2) has a unique generalized solution.

Setting $y = \{0, y_2\}$, where $y_2 \in D(|M|^{1/2})$, in identity (3.6), we find that the function $|M(a)|^{-1/2} x_1(t)$ is weakly differentiable and

$$\frac{d}{dt}[|M(a)|^{-1/2} x_1(t)] = [|M(t)|^{1/2} |M(a)|^{-1/2}]^{*} (\operatorname{sign} M(t)) |M(t)|^{1/2} x_2(t). \tag{6.7}$$

If we take $y = \{v, 0\}$, where $v \in D(|P_0|^{1/2})$, in (3.6), then with the help of (6.7) we will find that

$$[Qu(a), v] + [M^{-1}(a)\,\dot{u}(a), v] - [Qu(t), v] - \left[(|M(a)|^{1/2} M^{-1}(t))_0^{*} \frac{d}{dt}(|M(a)|^{-1/2} u(t))\right] =$$
$$= \int_a^t \{[|P_0(s)|^{1/2} u(s), |P_0(s)|^{1/2} v] \operatorname{sign} P_0(s) + [P_1(s) u(s), v]\} ds. \tag{6.8}$$

Here, we have used the abbreviation $u(t) = x_1(t)$.

It is clear from what has been said above that the function u(t) possesses the properties listed in the following definition.

Definition 3. Any measurable function that possesses the properties listed below will be called the *generalized solution* of Eq. (6.1).

1. The function $|P_0(a)|^{1/2} u(t)$ has meaning and is weakly continuous with respect to t.
2. The weak derivative $\frac{d}{dt}|M(a)|^{-1/2} u(t)$ exists and is a weakly continuous function.

3. For any element v belonging to $D(|P_0|^{1/2})$, the function u(t) satisfies the integral identity (6.8) and by weak continuity the first of the initial conditions (6.2).

It can be easily shown that if u(t) possesses the above properties, then function (6.4) is a generalized solution of problem (3.1)-(3.2).

Thus, we have proved

THEOREM 8. Let Conditions IX, X, and XI be satisfied. If $u(a) \in D(|P_0|^{1/2})$ and $\dot{u}(a) \in R$, then Eq. (6.1) with the initial conditions (6.2) has a unique generalized solution.

Let us now consider in space R the equation

$$\frac{d}{dt}\left[M(t)\frac{d}{dt}u(t)\right] + Q\frac{d}{dt}u(t) + P(t)u(t) = 0 \tag{6.9}$$

with the initial conditions (6.2). Here, the operator P(t) is given by (6.3) and, in addition to Conditions IX and X, the operators M(t), Q, $P_0(t)$, and $P_1(t)$ satisfy the following:

XII. The operator $|M(t)|^{1/2}|P_0(t)|^{-1/2}$ has meaning and is uniformly bounded with respect to $t \in [a, b)$.

XIII. For almost all t, the bounded operator $|M(t)|^{1/2}P_1(t) \times |P_0(a)|^{-1/2}$, which is measurable, has meaning for the symmetric operator $P_1(t)$ and its norm is summable on segment [a, b].

With the above assumptions, Eq. (6.9) has a generalized solution provided that $u(a) \in D(|P_0|^{1/2})$ and $\dot{u}(a) \in D(|M|^{1/2})$.

Definition 4. A measurable function u(t) will be said to be a generalized solution of Eq. (6.9) if it possesses the following properties:

1. The function $|P_0(a)|^{1/2}u(t)$ has meaning and is weakly continuous with respect to t.
2. The weak derivative $\frac{d}{dt}|M(a)|^{1/2}u(t)$ exists and is a weakly continuous function of t.

3. For any element v belonging to $D(|P_0|^{1/2})$, the function u(t) satisfies the integral identity

$$[Qu(a), v] + [|M(a)|^{1/2}\dot{u}(a), |M(a)|^{1/2}v] \operatorname{sign} Ma - [Qu(t), v] - \left[|M(t)|^{1/2}\frac{du(t)}{dt}, |M(t)|^{1/2}v\right] \operatorname{sign} M(t) =$$

$$= \int_a^t \{[|P_0(s)|^{1/2}u(s), |P_0(s)|^{1/2}v] \operatorname{sign} P_0(s) + [P_1(s)u(s), v]\} as, \tag{6.10}$$

and by continuity the first of the initial conditions (6.2).

THEOREM 9. Let Conditions IX, X, XII, and XIII be satisfied. If $u(a) \in D(|P_0|^{1/2})$ and $\dot{u}(a) \in D(|M|^{1/2})$, then Eq. (6.9) with initial conditions (6.2) has a unique generalized solution.

Proof. The function $v(t) = |M(a)|^{1/2}u(t)$ satisfies an equation of type (6.1) with appropriate initial conditions. This equation satisfies the conditions of Theorem 8 from which the assertions of Theorem 9 can be derived.

If operator $P_0(t)$ is positive for all t, then, setting M(t) = I and Q = 0 in (6.1), we will obtain the hyperbolic equation

$$\frac{d^2u(t)}{dt^2} + P(t)\,u(t) = 0, \tag{6.11}$$

for which Theorem 8 with M = I and Q = 0 is valid.

In view of the importance of Eq. (6.11) for various applications, let us formulate the above results as a theorem with the following conditions:

XIV. $P_0(t)$ is a self-adjoint positive definite operator, i.e., $P_0^*(t) = P_0(t) \geq \alpha I$ (where $\alpha > 0$ is a constant).

XV. The domain $D(P_0^{1/2})$ of definition of the positive square root $P_0^{1/2}(t)$ is independent of $t \in [a, b)$.

XVI. The interval $[a, b)$ can be subidivided by means of points $a = t_0 < t_1 < \ldots < t_n = b$ into a finite number of half-open intervals $[t_{k-1}, t_k)$, on each of which the inequality

$$\| P_0^{1/2}(t) P_0^{1/2}(s) \| \leqslant 1 + \text{const}\,|t - s|$$

holds for $t, s \in [t_{k-1}, t_k)$.

XVII. A bounded measurable operator $P_1(t) P_0^{1/2}(a)$ is defined for the symmetric operator $P_1(t)$ for almost all t and its norm is Lebesgue integrable on $[a, b)$.

THEOREM 10. Let the operator P(t) be represented as the sum (6.3), both terms of which satisfy requirements XIV-XVII. If $u(a) \in D(P_0^{1/2})$ and $\dot{u}(a) \in R$, then Eq. (6.11) with initial conditions (6.2) has a unique generalized solution which is to be understood in the sense of Definition 3 with M = I and Q = 0.

On the basis of Theorem 7, the properties of the generalized solution will improve with improving properties of the initial data and improving conditions of smoothness of the coefficients of Eqs. (6.1) and (6.9). In view of the complexity of the corresponding conditions, this result will be formulated below only for the hyperbolic equation (6.11).

XVIII. $D(P_0)$, the domain of definition of the operator $P_0(t)$, is independent of t. The interval $[a, b)$ can be subdivided by means of points $a = t_0 < t_1 < \ldots < t_n = b$ into a finite number of half-open intervals $[t_{k-1}, t_k)$, on each of which the inequality

$$\| P_0(t) P_0^{-1}(s) \| \leqslant 1 + \text{const}\,|t - s|$$

holds for $t, s \in [t_{k-1}, t_k)$.

XIX. The operator $P_0^{1/2}(a) P_1(t) P_0^{-1}(a)$, bounded for almost all t and measurable on $[a, b)$, has meaning for the symmetric operator $P_1(t)$ and its norm is summable.

THEOREM 11. Let Conditions XIV, XVIII, and XIX be satisfied. If $u(a) \in D(P_0)$ and $\dot{u}(a) \in D(P_0^{1/2})$, then the generalized solution* of Eq. (6.11) with the initial conditions (6.2) possesses the following properties:

*With the above assumptions, all conditions of Theorem 10 are satisfied and, consequently, the generalized solution exists.

1. The weakly continuous functions $P_0(a)u(t)$ and $P_0^{1/2}\dot{u}(t)$, where $\dot{u}(t)$ is the strong derivative of $u(t)$, have meaning.

2. For almost all t, the function $u(t)$ has a strong second derivative and satisfies Eq. (6.11).

Proof. The coefficients of Eq. (3.1), to which Eq. (6.11) can be reduced by the change of variable indicated above, clearly satisfy the conditions of Theorem 7 when the coefficients of Eq. (6.11) satisfy Conditions XIV, XV, XVI, XVIII, and XIX. It directly follows from this that the assertions of Theorem 11 hold when the additional restrictions XV and XVI are imposed. Using Lemma 3 of [9] as in Section 5, we can prove that conditions XV and XVI are a consequence of the other conditions of Theorem 11.

§ 7. Nonlinear Equations

The assertions of Sections 2-6 can be extended in an obvious manner to Eqs. (3.1), (6.1), (6.9), and (6.11) with nonsymmetric operators $H_1(t)$ and $P_1(t)$. We can introduce a nonhomogeneous term or even a nonlinearity into the above equations.

As a concrete example, let us consider the equation

$$J\frac{dx}{dt} = H_0(t)x + f(x, t) \tag{7.1}$$

with the initial condition

$$x(t)|_{t=a} = x_0 \in D(H^{1/2}(a)). \tag{7.2}$$

We will assume that the operator $H_0(t)$ satisfies conditions I-III.

The function $f(x, t)$ is assumed to be defined for all $t \in [a, b]$ and for all $x \in W$, belonging to the sphere $\|x - x_0\| < \rho$ of a sufficiently large radius ρ. The function $f(x, t)$ is strongly continuous with respect to the set of variables x, t and satisfies the inequality

$$\|H_0^{1/2}(0)J^{-1}[f(H_0^{-1/2}(0)z_1, t) - f(H_0^{-1/2}(0)z_2, t)]\| \leqslant q\|z_1 - z_2\|,$$

where q is a constant, for all z_1, z_2 belonging to the sphere $\|z - x_0\| < \rho$.

With the above assumptions, the Cauchy problem (7.1)-(7.2) has a unique generalized solution, i.e., there exists a unique function x(t) which satisfies both the integral identity

$$(J[x(t) - x_0], y) = \int_a^t [(H_0^{1/2}(s)x(s), H_0^{1/2}(s)x(s)) + (f(x, t), y)]\,ds$$

for any element $y \in D(H_0^{1/2}(a))$ and the inequality

$$\|H_0^{1/2}(a)x(t)\| \leqslant \mathrm{const}\|H_0^{1/2}(a)x(a).$$

As applied, for example, to the second order equation

$$\ddot{u} + P_0(t)u + g(u, t) = 0 \tag{7.3}$$

the last assertion can be reformulated as follows.

We solve Eq. (7.3) with the initial condition

$$u(a) \in D(P_0^{1/2}(a)), \quad \dot{u}(a) \in R. \tag{7.4}$$

Let us assume that conditions X are satisfied for operator $P_0(t)$. The function g(u, t) is assumed to be given for all $t \in [a, b]$ and for all the $u \in R$, that belong to the sphere $\| u - u(a) \|_R < \rho$ of sufficiently large radius ρ. The function g(u, t) is continuous with respect to the set of variables u, t and satisfies the inequality

$$\| P_0^{1/2}(0) \left[g\left(P_0^{1/2}(a) v_1,\ t\right)\right] - \left[g\left(P_0^{1/2}(a) v_2,\ t\right)\right] \| \leqslant q \| v_1 - v_2 \|_R,$$

where q is a constant, for any v_1 and v_2 that belong to the sphere $\| v - u(a) \|_R < \rho$.

With the above assumptions, problem (7.3)-(7.4) has a unique generalized solution, i.e., the corresponding integral identity and the inequality

$$\| P_0^{1/2}(a)\, u(t) \|_R + \| \dot{u}(t) \|_R \leqslant \text{const} \left[\| P_0^{1/2}(a)\, u(a) \|_R + \| \dot{u}(a) \|_R \right]$$

are satisfied by u(t).

Literature Cited

1. Yu. L. Daletskii, Uspekhi Matem. Nauk, Vol. 17, No. 5 (107) (1962).
2. V. V. Nemytskii, M. M. Vainberg, and R. S. Gusarova, Progress in Mathematics: Mathematical Analysis [in Russian] (1964 and 1965).
3. V. I. Derguzov and V. A. Yakubovich, Dokl. Akad. Nauk SSSR, Vol. 151, No. 6 (1963).
4. T. Kato, "Integration of the equation of evolution in Banach space," J. Math. Soc. Japan, Vol. 5 (1953).
5. O. A. Ladyzhenskaya, Matem. Sbornik, Vol. 39 (81), No. 4 (1956).
6. M. I. Vishik, Matem. Sbornik, Vol. 39 (81), No. 1 (1956).
7. E. Hille and R. Phillips, Functional Analysis and Semigroups [Russian translation], Izd. "Mir," Moscow (1962).
8. L. V. Kantorovich and G. P. Akilov, Functional Analysis in Normed Spaces [in Russian], GIFML, Moscow-Leningrad (1959).
9. V. I. Derguzov, Matem. Sbornik, Vol. 63 (105), No. 4 (1964).
10. V. A. Yakubov, Vestnik Leningradsk. Gos. Univ., No. 12, Issue 3 (1958).

RESONANCE IN THE OSCILLATION OF LINEAR SYSTEMS UNDER THE ACTION OF AN ALMOST PERIODIC PARAMETRIC PERTURBATION

V. N. Fomin

The phenomenon of parametric resonance is well known – the study of the conditions for its occurrence is one of the central problems of the theory of stability of elastic systems [1-4]. The case of periodic loading has been studied in sufficient detail [1-10]: Formulas for the effective calculation of the resonance frequencies of the perturbation, and the determination of the domains of dynamic instability adjoining them have been obtained and other problems of the theory of parametric resonance have been solved. The present article represents an attempt to develop the analogous theory in the case of systems subject to the action of an almost-periodic parametric perturbation. The latter is a more substantial case inasmuch as there are effects which do not have an analog in the periodic case. The existence of these effects prevents us from developing the theory as fully as could be expected at first sight. However, the basic results of the theory of parametric resonance can be extended to systems of the type under investigation.

The article consists of four parts, of which only two are contained in the present volume. The first part is devoted to a description of the class of operator equations being studied and contains a discussion of the potentialities of the method of averaging for the determination of the index of exponential growth of solutions. The formulation of the problem of parametric resonance in linear almost-conservative systems is given in the second part and formulas for the determination in parameter space of "essential-instability" sets adjoining the resonance frequencies are derived. In the third part, it is proposed to extend the formulation of the problem of parametric resonance to a wider class of equations including, in particular, systems with "friction." Finally, the fourth part will be devoted to a study of oscillatory systems with the aim of verifying that the conditions imposed on the class of systems being studied are satisfied.

It should be noted that although systems with an infinite number of degrees of freedom described by partial differential equations are considered in this article, the results obviously remain valid for the corresponding systems with a finite number of degrees of freedom and, moreover, most of them in this case are apparently new.

PART I

OPERATOR EQUATIONS WITH ALMOST-PERIODIC COEFFICIENTS AND THE METHOD OF AVERAGING

§ 1. Quasi-Hamiltonian Equations and the Index of Exponential Growth of Their Solutions

Let us consider in Hilbert space **H** the differential equation

$$J\frac{dx}{dt}=[H_0+\varepsilon H(t,\ \omega)]\,x.^* \tag{1.1}$$

Here, J is an anti-Hermitian operator (i.e., $J^* = -J$) that is bounded together with its inverse and H_0 is a positive definite self-adjoint operator (i.e., $H_0 \geq \alpha I$, where $\alpha > 0$ and I is the unit operator in **H**). The operator H_0^{-1} (the inverse of H_0) is assumed to be completely continuous and to commute with J (i.e., $JH_0^{-1} = H_0^{-1}J$). The operator function $H(t, \omega)$, whose first argument is a scalar, $-\infty < t < \infty$, and second argument ω is a vector, can be represented as a finite sum

$$H(t,\ \omega)=\sum_k e^{i(\omega,k)t}H^{(k)}, \tag{1.2}$$

in which ω is an n-component real vector called the frequency vector, k is a vector with integer components, and $(\omega, k) = \sum_{j=1}^{n}\omega_j k_j$. The linear operators $H^{(k)}$ in (1.2) are assumed to be bounded and are called the Fourier coefficients of the operator function $H(t, \omega)$. Equation (1.1) with $H(t, \omega) \equiv 0$, i.e., the equation

$$J\frac{dx}{dt}=H_0x \tag{1.3}$$

will be called the unperturbed equation, whereas Eq. (1.1) will be called the perturbed equation. The positive parameter ε in Eq. (1.1) is called the perturbation amplitude; it is assumed that it can be arbitrarily small. Let us agree to use the term solution of Eq. (1.1) to denote the function x(t) with values in **H** such that the function $H_0^{-1}Jx(t)$ has a strong derivative which satisfies the equation

$$\frac{d\left(H_0^{-1}Jx\right)}{dt}=[I+\varepsilon H_0^{-1}H(t,\omega)]\,x. \tag{1.4}$$

It is not difficult to see that in view of the assumptions introduced above concerning the operator coefficients, for every element x_0 belonging to **H** there exists a unique solution x(t) of Eq. (1.1) satisfying the condition $x(0) = x_0$. Let X(t) denote the resolving operator (operator solution) of Eq. (1.1) defined in terms of the solution x(t) and the initial condition $x(0) = x_0$ by

* The space **H** can also be finite-dimensional. In this case, Eq. (1.1) represents in essence a system of ordinary differential equations whose order is equal to the number of dimensions of **H**. We will say that Eq. (1.1) in this case describes a system with a finite number of degrees of freedom. By analogy, when **H** is an infinite-dimensional space, we will say that Eq. (1.1) describes a system with an infinite number of degrees of freedom. Such systems are usually described by partial differential equations, so that they are sometimes called systems with distributed parameters. In the following, we do not exclude the case of a finite-dimensional space **H**. In places where the fact that **H** is infinite-dimensional is significant, an explicit statement to this effect will be given.

$$X(t)x_0 = x(t). \tag{1.5}$$

The operator function X(t) satisfies the integral equation

$$X(t) = e^{J^{-1}H_0t}[I + \varepsilon \int_0^t e^{-J^{-1}H_0s} J^{-1}H(s, \omega) X(s)\, ds], \tag{1.6}$$

where the integral is to be understood in the sense of Bochner [11]. Since the operator solution $e^{J^{-1}H_0t}$ of the unperturbed Eq. (1.3) is a unitary operator for all t, $-\infty < t < \infty$, Eq. (1.6) implies that the operator X(t) and its inverse $X^{-1}(t)$ are bounded for all t.

An equation of type (1.1) whose coefficients satisfy the conditions listed above will be called a quasi-Hamiltonian equation and in the special case when the values of the function H(t, ω) are symmetric operators, it will be called a Hamiltonian equation. In the following, we will pay particular attention to Hamiltonian equations.

An important characteristic of the equations being considered is the quantity $\rho = \rho(\varepsilon, \omega)$ which is called the *index of exponential growth* (or the *leading exponent*) of solutions of Eq. (1.1). It is defined by

$$\rho(\varepsilon, \omega) = \overline{\lim_{t\to\infty}} \frac{\ln \| X(t) \|}{t}, \tag{1.7}$$

where ε is the perturbation amplitude, ln the principal branch of the natural logarithm, and $\| X \|$ the norm of the operator X in **H**.

It follows from formula (1.7) [13] that for an arbitrary positive number μ there exists a solution $\tilde{x}(t)$ of Eq. (1.1) such that

$$\overline{\lim_{t\to\infty}} \| \tilde{x}(t) \| e^{[\mu-\rho(\varepsilon,\omega)]t} = \infty,$$

whereas any solution x(t) satisfies the relation

$$\overline{\lim_{t\to\infty}} \| x(t) \| e^{[-\mu-\rho(\varepsilon,\omega)]t} = 0.$$

Hence, if $\rho(\varepsilon, \omega) < 0$, then all solutions of Eq. (1.1) tend to zero as $t \to \infty$ (asymptotic stability). When $\rho(\varepsilon, \omega) > 0$, the equation is unstable and the oscillations grow exponentially. It should be noted that the Hamiltonian equation can never be asymptotically stable since we always have $\rho(\varepsilon, \omega) \geq 0$.†

† Indeed, the operator solution X(t) of a Hamiltonian equation satisfies the relation $X^*(t)JX(t) = J$ which is a statement of its J-unitary property. This relation yields the inequality

$$C_1 \|X(t)\| \leqslant \|X^{-1}(t)\| \leqslant C_2 \| X(t) \|,$$

where C_1 and C_2 are positive constants. It follows from this that we have

$$\overline{\lim_{t\to\infty}} \frac{\ln \| X(t) \|}{t} = \overline{\lim_{t\to\infty}} \frac{\ln \| X^{-1}(t) \|}{t}.$$

On the other hand, we always have $\| X(t) \| \cdot \| X^{-1}(t) \| \geqslant 1$, so that

$$\overline{\lim_{t\to\infty}} \frac{\ln \| X(t) \|}{t} + \overline{\lim_{t\to\infty}} \frac{\ln \| X^{-1}(t) \|}{t} \geqslant 0,$$

i.e., $\overline{\lim_{t\to\infty}} t^{-1} \ln \| X(t) \| \geqslant 0$, which was originally asserted.

In many cases of practical interest, it is possible to show that the function $\rho(\varepsilon, \omega)$ is of the form

$$\rho(\varepsilon, \omega) = \varepsilon \rho_1(\omega) + o(\varepsilon),$$

where ρ_1 is independent of ε. An important problem is the derivation of effective formulas for the determination of ρ_1; it is this quantity that usually governs the boundedness or unboundedness of the oscillations described by the perturbed equation for small ε. Moreover, the coefficients of Eq. (1.1) frequently depend on one or more parameters (for example, we can consider the frequency vector ω as a set of such parameters). In this case, the ability to calculate ρ_1 effectively allows us to identify in parameter space the most "dangerous" sets, i.e., sets whose elements are parameters that give the most rapidly growing oscillations of the system.

Similar problems have been studied in [9-10] in the case of a quasi-Hamiltonian equation on the assumption that the perturbations are periodic in time.*

In this case, the problem of the determination of the exponential-growth index of the solution was reduced to the problem of the determination of the spectral limit for the boundary value problem defined by Eq. (1.1) and periodic boundary conditions. A useful feature of the latter problem is that the spectral limit points with largest real parts were, as a rule, isolated eigenvalues. This made possible the formulation of a perturbation method which provides an effective procedure for the approximate calculation of the exponential-growth index of the solutions of the equation for sufficiently small values of this index and ε. It is possible to formulate such a boundary-value problem in the case of almost-periodic operator functions, while the study of operators generated by Eq. (1.1) in the space of almost-periodic functions introduces great difficulties in the problem of the determination of the spectral limit of these operators.

In the present article we have adopted another method which is simpler and, in a definite sense, more general than the above. It has proved itself in studies of systems of ordinary differential equations and has become widely known as the method of averaging.

It should be noted straight away that the method of averaging is usually used for the investigation of the behavior of solutions on a large (of the order of $1/\varepsilon$) interval of variation of t, whereas the problem of the determination of the function $\rho(\varepsilon, \omega)$ requires the analysis of the solutions of the equation on an infinite time interval. Therefore, we must make a special investigation of the possibility of neglecting small terms in the averaged equation.

Similar problems for systems with a finite number of degrees of freedom have been considered in [15-16] whose most significant feature is the use of a formal analog of the Floquet–Lyapunov theorem [15].

*It is not difficult to see that the transformation of Eq. (1.1) into

$$i \frac{d}{dt}(Fy) = [I + \varepsilon H_1(t)]\, y, \qquad (*)$$

the equation considered in [9-10], can be achieved by means of the following changes of variables $y = H_0^{-1/2}x$, $H_1(t) = H_0^{-1/2}H(t)H_0^{-1/2}$, and $F = -iH_0^{-1/2}JH_0^{-1/2}$. The inverse transformation can also be easily achieved. From the point of view of practical applications, it is immaterial whether we take quasi-Hamiltonian equations in the form of (1.1) or (*). However, the difference between the two forms is of importance in theoretical investigations because of the unboundedness of the transformation realizing the change from Eq. (1.1) to (*) (see [14] on this topic). In the present article the form (1.1) was found to be the most convenient.

§ 2. The Method of Averaging

Speaking in a somewhat general manner, we can say that the method consists in the discovery of a suitable change of variable which systematically simplifies the original system. This idea was already used by the founders of celestial mechanics (Lagrange, Poincaré) and was extensively developed by N. M. Krylov and N. N. Bogolyubov [17] (see also [18]).

The essence of the method in the case being considered can be described as follows.

By successive integrations of the integral equation (1.6), we can find X(t) in the form of a series in powers of ε converging uniformly when t varies within any finite interval

$$X(t) = e^{J^{-1}H_0 t}\left[J + \varepsilon\int_0^t e^{-J^{-1}H_0 s}J^{-1}H(s,\omega)\,e^{J^{-1}H_0 s}ds + \varepsilon^2\Phi(\varepsilon,t)\right], \tag{2.1}$$

where $\Phi(\varepsilon, t)$ is an operator function that is entire with respect to ε for any finite value of t. Although this series is not convenient for a study of the behavior of the solution over the whole of the numerical axis, we can attempt to use segments of it to perform a change of variable that will simplify the original equation. For example, the transformation

$$x(t) = e^{J^{-1}H_0 t}\left[I + \varepsilon\int_0^t B(s)\,ds\right]z(t), \tag{2.2}$$

where

$$B(t) = e^{-J^{-1}H_0 t}J^{-1}H(t,\omega)\,e^{J^{-1}H_0 t} \tag{2.3}$$

introduces the function z(t) satisfying the equation

$$\left[I + \varepsilon\int_0^t B(s)\,ds\right]\frac{dz}{dt} = \varepsilon^2 B(t)\int_0^t B(s)\,ds\,z(t). \tag{2.4}$$

If Eq. (2.4) is solved for dz/dt, we will obtain an equation that is simpler than the original one. However, such a transformation may be found to be invalid because of the possibility of the introduction of "secular" terms in transformation (2.2). In fact, for large values of t the function $\int_0^t B(s)\,ds$, with B(t) defined by (2.3) may give a term proportional to t which will not allow us to solve (2.3) for dz/dt at sufficiently large t however small the parameter ε. Therefore, it is usual to eliminate the secular terms and for this purpose the part proportional to t is subtracted from the integral in transformation (2.2). Namely, the integral is represented as

$$\int_0^t B(s)\,ds = B_0 t + \Psi(t), \tag{2.5}$$

where B_0 is the average value of the operator function B(t) and expression (2.2) is replaced by

$$x(t) = e^{J^{-1}H_0 t}[I + \varepsilon\Psi(t)]\,z(t). \tag{2.6}$$

The function z(t) now satisfies the equation

$$[I + \varepsilon\Psi(t)]\frac{dz}{dt} = \varepsilon B_0 z + \varepsilon^2 B(t)\,\Psi(t)\,z \tag{2.7}$$

instead of (2.4). In some cases, the operator function $\Psi(t)$ is found to be uniformly bounded with respect to t and Eq. (2.7) can then be written as

$$\frac{dz}{dt} = \varepsilon B_0 z + \varepsilon^2 B_1(\varepsilon, t) z, \tag{2.8}$$

where the function $B_1(\varepsilon, t) = [I + \varepsilon\Psi(t)]^{-1}[-\Psi(t) B_0 + B(t)\Psi(t)]$ is uniformly bounded with respect to t, with $-\infty < t < \infty$, for sufficiently small ε. Equation (2.8) is called the *averaged equation*. The use of the averaged equation allows us to introduce a "slow time" $\tau = \varepsilon t$ which, in turn, facilitates the investigation of the behavior of the solution as $t \to \infty$. Moreover, it follows from substitution (2.6) that the indices of the exponential growth of the solutions of the original Eq. (1.1) and of the averaged equation (2.8) are the same. Thus, the original problem of the approximate determination of $\rho(\varepsilon, \omega)$ for Eq. (1.1) is reduced in this case to the simpler problem of the determination of the analogous quantity for Eq. (2.9). If the spectrum of the operator B_0 is strictly contained within the left half-plane, then Eq. (2.8) is asymptotically stable for sufficiently small ε [i.e., $\rho(\varepsilon, \omega) < 0$]. If the operator B_0 has points of the spectrum in the right half-plane, then Eq. (2.8) is unstable [i.e., $\rho(\varepsilon, \omega) > 0$]. Such results are obtained comparatively simply for Eq. (2.8) with the help of Lyapunov's second method. Our problem is the determination of more refined estimates of the exponential-growth index $\rho(\varepsilon, \omega)$ for sufficiently small values of ε.

§ 3. Almost-Periodic Functions with Values in Hilbert Space

In this section we collect together information concerning almost-periodic functions which we will need to discuss the possibility of the transition from Eq. (1.1) to the averaged equation (2.8) [in particular, we give conditions for the transformation (2.6) to be uniformly bounded with respect to t, $t \geq 0$].

Let **R** denote the linear manifold of trigonometric polynomials of the form

$$y(t) = \sum_l x_l e^{i\mu_l t}, \tag{3.1}$$

where x_l are elements of a separable Hilbert space **H**, the μ_l are real numbers, and the sum contains only a finite number of terms. Let **C** denote the closure of set **R** in a metric generated by the form

$$|y(t)|_C \equiv \sup_{-\infty < t < \infty} \|y(t)\|, \tag{3.2}$$

where $\|y\| = (y, y)^{1/2}$ is the form of element y in space **H**. The elements of space **C** possess many of the properties characteristic of the usual uniformly almost-periodic functions [19]. The properties that will be required below are formulated as the following assertions.

LEMMA 1. *For any function $y(t) \in$ **C** there exists an average value $\bar{y}$ defined as the strong limit*

$$\bar{y} = \lim_{t \to \infty} \frac{1}{2t} \int_{-t}^{t} y(t)\, dt.$$

Proof. Let $y_n(t)$ be trigonometric polynomials approximating y(t). It is obvious that the functions $y_n(t)$ possess the required limits which we denote by $\bar{y}_n$. Let us show that we have $\|\bar{y}_n - \bar{y}_m\| \xrightarrow[n, m \to \infty]{} 0$. Indeed, we have

$$\|y_n - y_m\| \leqslant \left\| y_n - \frac{1}{2t}\int_{-t}^{t} y_n(s)\,ds \right\| + \left\| y_m - \frac{1}{2t}\int_{-t}^{t} y_m(s)\,ds \right\| + \frac{1}{2t}\int_{-t}^{t} \|y_n(s) - y_m(s)\|\,ds.$$

Since the convergence of $y_n(t)$ to $y(t)$ is uniform with respect to t, with $-\infty < t < \infty$, it follows that for sufficiently large n, m, and t the right-hand side of the above inequality can be made as small as we please. Thus, $\|y_n - y_m\| \to 0$ as $n, m \to \infty$ and, consequently, the sequence $\{y_n\}$ has a limit which will be denoted by y:

$$y = \lim_{n\to\infty} y_n.$$

It follows from the inequality

$$\left\| y - \frac{1}{2t}\int_{-t}^{t} y(s)\,ds \right\| \leqslant \|y - y_n\| + \left\| y_n - \frac{1}{2t}\int_{-t}^{t} y_n(s)\,ds \right\| + \frac{1}{2t}\left\| \int_{-t}^{t} y_n(s) - y(s) \right\| ds$$

that

$$\lim_{t\to\infty} \left\| y - \frac{1}{2t}\int_{-t}^{t} y(s)\,ds \right\| = 0,$$

which was to be proved.

It follows from Lemma 1 that the following strong limit exists for any real number λ and function $y(t) \in \mathbf{C}$:

$$\lim_{t\to\infty} \frac{1}{2t}\int_{-t}^{t} y(s)\,e^{-i\lambda s}\,ds = y_\lambda. \tag{3.3}$$

Definition 1. The number λ is called the Fourier index of the function y(t) if the element y_λ of **H** defined by formula (3.3) is different from zero. In this case, the element y_λ itself is called the Fourier coefficient of y(t). The closure of the set of all Fourier indices of y(t) contained in an interval $[a, b]$ is called the Fourier spectrum of y(t) in this interval.

It can be easily shown that, as usual, every function $y(t) \in \mathbf{C}$ has at most a countable number of Fourier indices and that it may be uniquely associated with the Fourier series

$$y(t) \sim \sum_l y_l e^{i\lambda_l t}, \tag{3.4}$$

where y_l are the coefficients and λ_l are the Fourier indices of y(t).

LEMMA 2. Parseval's equality holds for functions $y(t) \in \mathbf{C}$, i.e.,

$$\lim_{t\to\infty} \frac{1}{2t}\int_{-t}^{t} \|y(s)\|^2 ds = \sum_l \|y_l\|^2.$$

The proof differs little from that usually given for the case of uniformly almost-periodic functions [19]. It follows from Parseval's equality that the Fourier series in turn uniquely defines the function $y(t) \in \mathbf{C}$.

LEMMA 3 (Favard's Theorem). Let us assume that the Fourier indices of the function $x(t) \in \mathbf{C}$ do not have condensation points in some neighborhood of zero. We then have

$$\int_0^t x(s)\,ds = x_0 t + y(t), \tag{3.5}$$

where $y(t) \in \mathbf{C}$ and x_0 is the average value of $x(t)$.

The proof coincides in many respects with the proof of the analogous assertion in the case of uniformly almost-periodic functions (see Theorem 1.12.1 of [19]) and consists of the following. By convention, the moduli of the Fourier indices of the function $x(t) - x_0$ are not less than some positive number α. Let us define a numerical function $\varphi(t)$ by

$$\varphi(\lambda) = \begin{cases} \frac{1}{i\alpha^2}\lambda, & 0 \leqslant |\lambda| < \alpha, \\ \frac{1}{i\lambda}, & |\lambda| > \alpha, \end{cases}$$

and let us denote its Fourier transform by $\psi(\lambda)$. The function

$$y(t) = \int_{-\infty}^{\infty} x(t+s)\,\psi(s)\,ds \tag{3.6}$$

then belongs to **C**. Indeed, if $x(t)$ is a trigonometric polynomial, then the function $y(t)$ defined by (3.6) will also be a trigonometric polynomial. Let $x_n(t)$ be the trigonometric polynomials approximating $x(t)$ and $y_n(t)$ be the corresponding functions obtained from formula (3.6). Proceeding to the limit under the integral sign, which is always possible because of the convergence of $x_n(t)$ to $x(t)$ uniformly with respect to t, we find that the sequence $\{y_n(t)\}$ converges to the function $y(t)$ uniformly with respect to t, so that $y(t)$ belongs to **C**. The conventional calculation of the Fourier coefficients y_λ of $y(t)$ shows that they are related to the Fourier coefficients x_λ of $x(t)$ as follows:

$$y_l = \frac{x_l}{i\lambda_l},$$

where λ_l are the Fourier indices of $x(t)$. It follows from this that the function $y(t)$ in formula (3.5) is of the same form as (3.6). The lemma has been proved. Let us now make use of the properties formulated above and proceed to the study of some of the properties of special operator functions acting in space **H**. In the following, we will be considering operators that possess some of the properties listed below.

Definition 2. We will say that an operator F possesses

Property S if F is a completely continuous symmetric operator with an inverse F^{-1} (bounded in the case of infinite-dimensional **H**);

Property S_δ if the operator F possesses property S and, in addition, on both the positive and negative halves of the real axis the operator has at least one lacuna (a gap in the points of regularity) whose length is not less than the given positive number δ;

Property S_∞ if the operator possesses Property S_δ for for an arbitrary δ.

It is obvious that Property S implies Property S_δ for some $\delta > 0$. As will be shown below, Properties S_δ and S_∞ allow us to "decompose" certain operators that are of interest to us in our investigation.

LEMMA 4. Let B(t) be an operator function of the form

$$B(t) = e^{iF^{-1}t} K(t) e^{-iF^{-1}t}, \tag{3.7}$$

where the operator F possesses Property S and K(t) is a trigonometric polynomial with bounded operator Fourier coefficients

$$K(t) = \sum_{l} e^{i\mu_l t} K_l. \tag{3.8}$$

Then, for any element x of space **H**, the function x(t) = B(t)x belongs to space **C**.

Proof. Since the sum in (3.8) contains a finite number of terms, it is sufficient only to consider the case when K(t) consists of one harmonic. For concreteness, let us consider the zero harmonic. In this case B(t) is of the form

$$B(t) = e^{iF^{-1}t} K e^{-iF^{-1}t},$$

where K is a bounded linear operator independent of t. Let E_λ be the spectral family [20] of the operator F^{-1} and let us set $P_\lambda = E_\lambda - E_{-\lambda}$. Let us consider the "truncated" operator function $B_\lambda(t) = P_\lambda B(t) P_\lambda$. Since for all finite λ the quantity P_λ is a projector on a finite-dimensional subspace, the function $x_\lambda(t) = B_\lambda(t)x$ is a trigonometric polynomial for any $x \in \mathbf{H}$ and, consequently, we have $x_\lambda(t) \in \mathbf{C}$. Let us now assume that the element $x \in \mathbf{H}$ is such that for all λ greater in modulus than a positive number M we have $P_\lambda x = x$. Then, for $|\lambda| \geq M$, $|\mu| \geq M$, we have

$$\| x_\lambda(t) - x_\mu(t) \| = \| (P_\lambda - P_\mu) B(t) x \| = \| (P_\lambda - P_\mu) K e^{-iF^{-1}t} x \| \leqslant \sum_{l} \| (P_\lambda - P_\mu) z_l \|,$$

where z_l are the Fourier coefficients of the function $z(t) = Ke^{-iF^{-1}t}x$. In view of the assumptions concerning the element x, the function z(t) is a trigonometric polynomial and, consequently, the summation on the right-hand side of the above inequality contains only a finite number of terms. Therefore, for $\lambda \to \infty$ and $\mu \to \infty$, we have

$$\sup_{-\infty < t < \infty} \| x_\lambda(t) - x_\mu(t) \| \to 0.$$

It follows from this that $x_\lambda(t) \xrightarrow{\mathbf{C}} x(t)$. The existence of the limit $\lim_{\lambda \to \infty} B_\lambda(t) x$ in **C** has therefore been established for elements x belonging to a set dense in **H**. Let us now assume that x is an arbitrary element of **H** and that $\{x_n\}$ is a sequence converging to x and satisfying the condition $P_\lambda x_n = x_n$ for $\lambda \geq M$, $M = M(n)$. We then have

$$\| B_\lambda(t) x - B_\mu(t) x \| \leqslant \| B_\lambda(t) x_n - B_\mu(t) x_n \| +$$
$$+ \| B_\lambda(t) - B_\mu(t) \| \cdot \| x - x_n \| \leqslant \| B_\lambda(t) x_n - B_\mu(t) x_n \| + 2 \| K \| \cdot \| x - x_n \|,$$

from which follows the convergence in itself of the sequence $B_\lambda(t)x$ uniformly with respect to $-\infty < t < \infty$. Since the space **C** is complete, the limit $\lim_{\lambda \to \infty} B_\lambda(t) x$ exists in **C** and, obviously, coincides with the function B(t)x. This completes the proof of the lemma.

It follows from Lemmas 1 and 4 that the following strong limit exists for functions B(t) satisfying the conditions of Lemma 4:

$$\lim_{t \to \infty} \frac{1}{2t} \int_{-t}^{t} B(t)\, dt \equiv B_0, \tag{3.9}$$

which we will call the average value of the operator function B(t).

LEMMA 5. Let the operator function B(t) be of the form of (3.7)-(3.8). Let λ_l denote the eigenvalues of the operator F^{-1} and let us assume that the set

$$\{\lambda_l - \lambda_m + \mu_s\}, \tag{3.10}$$

where the indices l, m, and s vary independently, does not have condensation points in some neighborhood of zero. Then, the operator function

$$\Psi(t) = \int_0^t e^{iF^{-1}s} K(s)\, e^{-iF^{-1}s}\, ds - B_0 t \tag{3.11}$$

is a bounded operator from **H** into **C**.*

Proof. Let x belong to **H** and let x(t) be the function defined by

$$x(t) = e^{iF^{-1}t} K(t)\, e^{-iF^{-1}t} x - B_0 x.$$

It is not difficult to see that the Fourier indices of x(t) belong to set (3.10) and, consequently, x(t) satisfies the conditions of Lemma 3. Relation (3.5) for x(t) now determines formula (3.11) for the operator function $\Psi(t)$. Let us now show that this function is bounded as an operator from **H** into **C**. It follows from formula (3.6) that the function $\Psi(t)$ can be represented as

$$\Psi(t)\, x = \int_{-\infty}^{\infty} \left[e^{iF^{-1}(s+t)} K(s+t)\, e^{-iF^{-1}(s+t)} - B_0\right] x \psi(s)\, ds,$$

which yields

$$\| \Psi(t)\, x \| \leqslant \sup_t (\| K(t) \| + \| B_0 \|) \int_{-\infty}^{\infty} |\psi(s)|\, ds \, \| x \| \leqslant \text{const} \| x \|,$$

or

$$|\Psi(t)\, x|_C \leqslant \text{const} \| x \|,$$

which was to be proved.

Remark. The conditions of Lemma 5 are automatically satisfied if the spectrum of the operator F^{-1} diverges in the sense that the quantity $|\lambda_l - \lambda_m|$ (with $\lambda_l \neq \lambda_m$) becomes arbitrarily large as l, m $\to \infty$.

The condition mentioned in the remark is a somewhat restrictive one and if it does not hold, the verification of the conditions of Lemma 5 is a difficult matter because it is associated with the investigation of the asymptotic distribution of the numbers λ_l as $l \to \infty$. In a number of cases of practical importance, it is possible to use the assertions given below to avoid the difficulties associated with the investigation of the distribution of the λ_l. The method consists in the decomposition of the operator F^{-1} into parts, it being finally found that the parts corresponding to λ_l with large indices are in a well-defined sense unimportant.

* In particular, if the Fourier spectrum of the operator function B(t) does not contain zero, then the integral $\int_0^t B(s)\, ds$ is a function bounded uniformly with respect to t.

LEMMA 6. Let us assume that the operator F possesses Property S_δ. Let q_1 and q_2 be the centers of the lacunas, whose widths are greater than δ, where $q_1 < 0$ and $q_2 > 0$. Let P_1, P_2, and P_3 denote the orthoprojectors of the operator F^{-1} on invariant subspaces, P_1 corresponding to the spectrum to the left of q_1, P_2 the spectrum between q_1 and q_2, and P_3 the spectrum to the right of q_2. Let the operator function B(t) be of the form of (3.7)-(3.8) with $\max_s |\mu_s| = \delta_1$. Then, for $\delta_1 < \delta$, the operator function

$$\Phi(t) = B(t) - \sum_{i=1}^{3} P_i B(t) P_i \tag{3.12}$$

is a bounded operator from **H** into **C**, the Fourier indices of the function $\Phi(t)$ being outside the interval $[\delta_1 - \delta, \delta - \delta_1]$.

Proof. The operator $\sum_{i=1}^{3} P_i B(t) P_i$ is a bounded operator from **H** into **C**. This is proved in the same way as the assertion of Lemma 4 concerning the operator B(t). The operator $\Phi(t)$ therefore also possesses this property. Let us prove the last part of the lemma. Let a_j be the normalized eigenelements of F^{-1} corresponding to the eigenvalues λ_j. We will assume that the numbers λ_j have been numbered in order of increasing $|\lambda_j|$, so that $\lambda_j \operatorname{sign} j > 0$. Let us consider the operator function B(t) in the basis formed by the elements $\{a_j\}$. The corresponding matrix elements are

$$(B(t) a_l, a_m) = \sum_s e^{i(\lambda_m - \lambda_l + \mu_s)t} (K_s a_l, a_m). \tag{3.13}$$

It follows from formulas (3.12) and (3.13) that the matrix elements $(\Phi(t) a_l, a_m)$ may be different from zero only if the numbers λ_l, λ_m belong to distinct regions of the spectrum, i.e., the relation $|\lambda_m - \lambda_l| > \delta$ must hold in any case for the nonzero elements $(\Phi(t) a_l, a_m)$. Consequently, for these values of l and m we have

$$|\lambda_m - \lambda_l + \mu_s| > \delta - \delta_1.$$

Inasmuch as the Fourier indices of the operator function $\Phi(t)$ are included in the set of the Fourier indices of its matrix elements $(\Phi(t) a_l, a_M)$, the lemma has been proved.

Remark. If the conditions of Lemma 6 are satisfied and the function $\Phi(t)$ is defined by formula (3.12), then it follows from Lemma 5 that the operator function $\Psi(t) = \int_0^t \Phi(s)\, ds$ is a bounded operator from **H** into **C**. By making various assertions concerning the operator function B(t), we can obtain a set of analogous assertions concerning the decomposition of it into two operator functions, one of which in a sense is simpler (for example, independent of t) and the other having Fourier indices outside some neighborhood of zero. Each such assertion allows us to reduce the original equation (1.1) to an equation whose principal part is the simpler operator function split-off from B(t). For example, assuming that the Fourier indices of the function B(t) that are situated in some neighborhood of zero have been separated from the other Fourier indices, we can immediately proceed to a new operator equation whose principal part $\tilde{B}(t)$ is the "projection" of B(t) on its Fourier spectrum situated in the above mentioned neighborhood of zero. It should be noted that we were able to perform this decomposition in Lemma 6 even if the Fourier indices of B(t) may be everywhere dense on the real axis.

§ 4. The Principal Theorem

This section will be devoted to a discussion of the correspondence between the indices of the exponential growth of the solutions of the equation

$$\frac{dz}{dt}=[\varepsilon B_0+\varepsilon^2 B(\varepsilon,\ t)]z \tag{4.1}$$

and the truncated equation

$$\frac{dz}{dt}=\varepsilon B_0 z. \tag{4.2}$$

for sufficiently small values of ε. It has been shown in the preceding sections that Eq. (1.1) can sometimes be reduced to (4.1) by means of a change of variable that is uniformly bounded with respect to t, $-\infty < t < \infty$. A more detailed investigation of the cases when such a reduction is possible will be made subsequently, while we assume now that the coefficients of Eq. (4.1) satisfy the following conditions:

1) The operator B_0 is linear and bounded, while the points of its spectra possessing the largest real parts are isolated eigenvalues corresponding to finite-dimensional radical spaces.

2) The function $B(\varepsilon, t)$ whose values are bounded linear operators in H is uniformly bounded with respect to t for values of $\varepsilon \geq 0$ that do not exceed a sufficiently small value ε_0.

It should be noted that these conditions are not restrictive for most applications.

The required correspondence between the exponential-growth indices will be established with the help of estimates of certain quadratic forms or Lyapunov functions. This method of the investigation of the stability of operator differential equations has been expounded by M. G. Krein [13]. The aim of the present section is the proof of the following assertion which serves as the justification of the method of averaging used in the present article.

THEOREM 1. Let us assume that the coefficients of Eq. (4.1) satisfy conditions 1) and 2) formulated above. Let $\rho(\varepsilon)$ and $\varepsilon\rho_1$ denote the indices of the exponential growth of solutions of Eqs. (4.1) and (4.2), respectively. Then, the following estimate holds for all ε belonging to the interval $[0, \varepsilon_0]$:

$$|\rho(\varepsilon)-\varepsilon\rho_1|<C\varepsilon^{1+\gamma}, \tag{4.3}$$

where C and γ are positive constants determined by the operator B_0 and depending on the structure of its radical subspaces. In particular, if the eigenvalues of the operator B_0 with the largest real parts correspond to elementary prime divisors, then $\gamma = 1$.

Remark 1. It follows from Theorem 1 that the function $\rho(\varepsilon)$ at zero depends continuously on the coefficients of Eq. (1.1) when the latter are varied in the metric of C (see [21]).

Remark 2. It is assumed in the theorem that the vector mentioned in the preceding section is fixed. The constant C in inequality (4.3) depends significantly on ω and this fact creates great difficulties in the study of the structure of the space of the parameters $\{\varepsilon, \omega\}$ of the system being considered.

Before we proceed to prove the theorem, let us give a number of auxiliary constructions and estimates.

Let Γ denote a line in the complex plane parallel to the imaginary axis and situated in the domain of regularity of the operator B_0. In view of the above assumptions, we can draw Γ

in such a manner that the eigenvalues of B_0 with the largest real parts will be situated in a given neighborhood of the line and will be separated by this line from the remaining spectrum. Performing the replacement $y = e^{\mu t}$ in Eqs. (4.1) and (4.2) and introducing the "slow" time $\tau = \varepsilon t$, we obtain the equations

$$\frac{dy}{d\tau} = B_\mu y + \varepsilon B\left(\frac{\tau}{\varepsilon}, \varepsilon\right) y, \tag{4.4}$$

$$\frac{dy}{d\tau} = B_\mu y, \tag{4.5}$$

where $B_\mu = B_0 - \mu I$. We will choose the number μ such that the line Γ for the operator B_μ coincides with the imaginary axis. Let Γ_1 and Γ_2 denote the smooth positively oriented contours situated in the right and left half-planes, respectively, and surrounding the spectrum of B_μ. Let us introduce the following operators:

$$P_j = \frac{1}{2\pi i} \oint_{\Gamma_j} (\zeta I - B_\mu)^{-1} d\zeta, \tag{4.6}$$

$$S_j = \frac{(-1)^j}{4\pi^2} \oint_{\Gamma_j^*} d\xi \oint_{\Gamma_j} d\zeta \frac{(\xi I - B_\mu^*)^{-1} (\zeta I - B_\mu)^{-1}}{\xi + \zeta}, \quad j = 1, 2. \tag{4.7}$$

Here, B_μ^* is the operator that is the adjoint of B_μ; Γ_1^* and Γ_2^* are positively oriented contours that are the complex conjugates of Γ_1 and Γ_2.

LEMMA 7. *There exist positive constants α_1, α_2, β_1, β_2 (where $\alpha_1 < \beta_1$ and $\alpha_2 < \beta_2$) which satisfy the following inequalities:*

$$\alpha_j \| P_j x \|^2 \leqslant (S_j x, x) \leqslant \beta_j \| P_j x \|^2, \quad j = 1, 2, \tag{4.8}$$

for any element x belonging to **H**. *Moreover, the following relations hold:*

$$B_\mu^* S_j + S_j B_\mu = (-1)^{j+1} P_j^* P_j, \quad j = 1, 2. \tag{4.9}$$

The proof of the lemma follows easily from the fact that it is possible to represent the operators S_j, j = 1, 2, as in [13]

$$\begin{aligned} S_1 &= \int_0^\infty P_1^* e^{-B_\mu^* t} e^{-B_\mu t} P_1 \, dt, \\ S_2 &= \int_0^\infty P_2^* e^{B_\mu^* t} e^{B_\mu t} P_2 \, dt. \end{aligned} \tag{4.10}$$

Indeed, if formulas (4.10) hold, then inequalities (4.8) are obvious and relations (4.9) can be verified immediately. Let us show for completeness how formula (4.7) can be transformed into (4.10), restricting ourselves to the case j = 1.

Since the contours Γ_1 and Γ_1^* are situated in the right half-plane, for $\xi \in \Gamma_1^*$ and $\zeta \in \Gamma_1$ we have

$$\int_0^\infty e^{-(\xi+\zeta)t} dt = \frac{1}{\xi + \zeta}.$$

Substituting the integral into formula (4.7) and making use of

$$P_1B_\mu = \frac{1}{2\pi i}\oint_{\Gamma_1} e^{-\zeta t}(\zeta I - B_\mu)^{-1}\,d\zeta,$$
$$P_1^* B_\mu^* = -\frac{1}{2\pi i}\oint_{\Gamma_1^*} e^{-\xi t}(\xi I - B_\mu^*)^{-1}\,d\xi,$$

we arrive at representation (4.10) for the operator S_1. The second formula (4.10) for S_2 can be obtained in an analogous manner.

LEMMA 8. a) Let us assume that the operator B_μ in Eq. (4.5) has spectral points in the right half-plane and that ε satisfies the inequalities

$$\left.\begin{aligned} \gamma_1(\varepsilon) \equiv \frac{1}{\beta_1} - 4\varepsilon\|S_1\|\cdot C(\varepsilon)\left[\frac{1}{\alpha_1}+\frac{1}{\alpha_2}\right] \geqslant 0,\\ 4\varepsilon\|S_1 - S_2\|C(\varepsilon) \leqslant 1, \end{aligned}\right\} \tag{4.11}$$

where $C(\varepsilon) = \sup_\tau \left\|B\left(\frac{\tau}{\varepsilon},\,\varepsilon\right)\right\|$ and α_1, α_2, β_1 are the positive constants of Lemma 7. Then, a solution $y(\tau)$ of Eq. (4.4) exists and satisfies the inequality

$$\|y(\tau)\|^2 \geqslant \frac{\alpha_1}{\beta_1}\|y(\sigma)\|^2 e^{\gamma_1(\varepsilon)(\tau-\sigma)}. \tag{4.12}$$

b) Let us now assume that the whole spectrum of the operator B_μ is situated in the left half-plane and that ε satisfies the inequality

$$\gamma_2(\varepsilon) \equiv 1 - 2\varepsilon\|S_2\|C(\varepsilon) \geqslant 0. \tag{4.13}$$

Then, any solution $y(\tau)$ of Eq. (4.4) satisfies the inequality

$$\|y(\tau)\|^2 \leqslant \frac{\beta_2}{\alpha_2}\|y(\sigma)\|^2 e^{-\gamma_2(\varepsilon)(\tau-\sigma)}.$$

Proof. a) Differentiating the quadratic forms $(S_j y, y)$, $j = 1, 2$, we have in view of Eq. (4.4)

$$\begin{aligned} \frac{d(S_1y,\,y)}{d\tau} &= \|P_1y\|^2 + 2\varepsilon\,\mathrm{Re}\left(S_1B\left(\frac{\tau}{\varepsilon},\,\varepsilon\right)y,\,y\right),\\ \frac{d(S_2y,\,y)}{d\tau} &= -\|P_2y\|^2 + 2\varepsilon\,\mathrm{Re}\left(S_2B\left(\frac{\tau}{\varepsilon},\,\varepsilon\right)y,\,y\right). \end{aligned} \tag{4.14}$$

Performing obvious transformations, we obtain the relation

$$\frac{d((S_1-S_2)\,y,\,y)}{d\tau} = \|P_1y\|^2 + \|P_2y\|^2 + 2\varepsilon\,\mathrm{Re}\left((S_1-S_2)\,B\left(\frac{\tau}{\varepsilon},\varepsilon\right)y,\,y\right) \geqslant (\|P_1y\|^2 + \|P_2y\|^2)\,[1 - 4\varepsilon\|S_1-S_2\|C(\varepsilon)],$$

which, in view of the second of inequalities (4.11), yields the inequality

$$(S_1y(\tau),\,y(\tau)) \geqslant (S_2y(\tau),\,y(\tau)) - (S_2y(\sigma),\,y(\sigma)). \tag{4.15}$$

Let us now assume that the initial element $y(\sigma)$ is such that we have $P_1y(\sigma) = y(\sigma)$ and, consequently, $P_2y(\sigma) = 0$. We then have $(S_2y(\sigma), y(\sigma)) = 0$ and inequality (4.15) can be rewritten as

$$(S_1y(\tau),\,y(\tau)) \geqslant (S_2y(\tau),\,y(\tau)). \tag{4.16}$$

Making use of (4.8) and (4.16), we obtain from the first equation of system (4.14)

$$\frac{d(S_1y,\ y)}{d\tau} \geqslant \| P_1 y \|^2 - 2\varepsilon \| S_1 \| C(\varepsilon) \| y \|^2 \geqslant \gamma_1(\varepsilon) \quad (S_1 y,\ y),$$

which after an integration and another application of inequality (4.8) leads to (4.12).

b) In this case the contour Γ_2 contains the whole spectrum of the operator B_μ and the operator P_2 defined by formula (4.6) is the unit operator, i.e., $P_2 = I$. From the second equation of (4.14) we have

$$\frac{d(S_2y,\ y)}{d\tau} \leqslant -\| P_2 y \|^2 + 2\varepsilon \| S_2 \| C(\varepsilon) \| y \|^2 \leqslant -(1 - 2\varepsilon \| S_2 \| C(\varepsilon)) \frac{(S_2y,\ y)}{\alpha_2},$$

from which we obtain $(S_2y(\tau),\ y(\tau)) \leqslant (S_2y(\sigma),\ y(\sigma))\, e^{-\gamma_2(\varepsilon)(\tau-\sigma)}$, which coincides with inequality (4.13) in view of (4.8).

Relation (4.12) shows that with the assumptions of the first part of the lemma, Eq. (4.4) has solutions growing exponentially as $\tau \to \infty$ whenever Eq. (4.5) has such solutions. Similarly, all solutions of Eq. (4.4) tend to zero exponentially when this property is possessed by Eq. (4.5). In order to obtain more accurate information concerning the exponential-growth index for the perturbed equation, let us consider the quantities $\gamma_1(\varepsilon)$ and $\gamma_2(\varepsilon)$ in greater detail.

We will assume that the contour Γ_1 is continuously deformed and approaches the spectrum encompassed by it. The smaller the area encompassed by Γ_1 and the closer the contour Γ_2 to the spectral points with the largest real values contained within it, the smaller will be the difference between the indices of exponential growth of the solutions of Eqs. (4.1) and (4.2). The conditions of Lemma 8 must of course continue to hold. This imposes certain restrictions on the possible variations in the properties of contours Γ_1 and Γ_2 and the quantity ε. We now proceed to an examination of these restrictions.

LEMMA 9. a) Let us assume that the operator B_μ in Eq. (4.5) has spectral points in the right half-plane and let Γ_1 be the contour appearing in formulas (4.6). Let δ be the distance between the contour Γ_1 and the spectrum of B_μ and let ν be the distance between Γ_1 and the imaginary axis. If the positive numbers ε_0, δ_0 are sufficiently small, then we can find a positive constant C depending on the coefficients of Eq. (4.4) and such that inequalities (4.11) will hold for all ε and δ satisfying the inequalities

$$\varepsilon < \varepsilon_0,\ \delta < \delta_0,\ \varepsilon C < \nu^2 \delta^{4(q-1)}, \tag{4.17}$$

where q is the biggest dimension of the Jordan blocks for the spectral points of the operator B_μ with the largest real parts.

b) Let us assume that the whole spectrum of the operator B_μ is situated in the left half-plane and that Γ_2 is a contour that contains this spectrum. Let δ denote the distance of the contour Γ_2 from the spectral points of B_μ with the largest real parts and ν the distance of Γ_2 from the imaginary axis. If the positive numbers ε_0 and δ_0 are sufficiently small, then we can find a positive constant C such that inequality (4.13) will hold for all ε, δ satisfying inequalities (4.17).

Proof. Let us estimate the numbers α_1 and β_1. It follows from the first formula of (4.10) that

$$(S_1y, y) = \int_0^\infty \|e^{-B_\mu t} P_1 y\|^2 dt = \int_0^\infty \|e^{-B_\mu P_1 t} P_1 y\|^2 dt. \tag{4.18}$$

For each $t \in [0, \infty)$ the operator $D(t) = e^{-B_\mu t} P_1$ provides a one-to-one mapping of the subspace P_1H into itself. The inverse operator $D^{-1}(t)$ is given by

$$D^{-1}(t) = e^{B_\mu P_1 t} P_1.$$

It is obvious that the norm of the operator $D^{-1}(t)$ can be estimated as

$$\|D^{-1}(t)\| = \sup_{y \in P_1 H} \frac{\|D^{-1}(t) y\|}{\|y\|} \leqslant e^{\|B_\mu P_1\| t}.$$

Taking into account that $D^{-1}(t)D(t)y = P_1y$, we obtain

$$\|D(t) y\| \geqslant \frac{\|P_1 y\|}{\|D^{-1}(t)\|} \geqslant e^{-\|P_1 B_\mu\| t} \|P_1 y\|.$$

Formula (4.18) now yields

$$(S_1 y, y) \geqslant \int_0^\infty e^{-2\|P_1 B_\mu\| t} \|P_1 y\|^2 dt = \frac{\|P_1 y\|^2}{2\|P_1 B_\mu\|}.$$

Consequently, we have

$$\alpha_1 \geqslant \frac{1}{2\|P_1 B_\mu\|}. \tag{4.19}$$

Let us obtain the upper bound to the form (S_1y, y). Since by assumption, the spectrum of B_μ is discrete in the right half-plane, the contour Γ_1 can be transformed into a set of circles of sufficiently small radius whose centers coincide with the eigenvalues of B_μ. Let $\{\overline{\lambda}_j\}$ denote the set of eigenvalues of B_μ with positive real parts. Then, as $\lambda \to \{\overline{\lambda}_j\}$ the resolvent $(\lambda I - B_\mu)^{-1}$ of B_μ satisfies the inequality

$$\|(\lambda I - B_\mu)^{-1}\| \leqslant \frac{C_1}{\delta^q}, \tag{4.20}$$

where δ and q are defined by the conditions of the lemma and C_1 is a constant depending on the operator B_μ. Making use of (4.20) we have for all sufficiently small δ

$$(S_1 y, y) \leqslant \frac{1}{4\pi^2} \int_{\Gamma_1} \|(\xi I - B_\mu)^{-1}\| \|P_1 y\| d\xi \int_{\Gamma_1} \frac{\|(\zeta I - B_\mu)^{-1}\| \|P_1 y\|}{|\zeta + \overline{\xi}|} d\zeta \leqslant \frac{C_2}{\nu \delta^{2q-2}} \|P_1 y\|^2,$$

where C_2 is a constant depending on B_μ and ν is the distance from the contour Γ_1 to the imaginary axis. Consequently, we have

$$\beta_1 \leqslant \frac{C_2}{\nu \delta^{2q-2}}. \tag{4.21}$$

Combining (4.19) and (4.17), we find that the first of inequalities (4.11) holds if

$$\frac{\varepsilon C}{\nu^2 \delta^{4(q-1)}} < 1,$$

where

$$C = 16C_2^2 \| P_1 B_\mu \| \cdot \sup_{0 \leqslant \varepsilon \leqslant \varepsilon_0} C(\varepsilon).$$

It is not difficult to see that when ν and δ are sufficiently small, if the first of inequalities (4.11) holds, then the second inequality will also hold. The first part of the lemma has been proved. The second part is proved in an analogous manner.

Indeed, inequality (4.13) holds when we have

$$2\varepsilon\beta_2 C(\varepsilon) < 1.$$

Inequality (4.21) can be obtained for β_2 and it follows that (4.13) will hold when we have

$$\frac{2\varepsilon C_2 C(\varepsilon)}{\nu\delta^{2(q-1)}} < 1.$$

It is obvious that the last inequality holds when conditions (4.17) are satisfied.

It is not difficult to prove Theorem 1 with the help of the inequalities of Lemmas 8 and 9. Indeed, let ρ_1 be the largest real part of the eigenvalues of the operator B_0. Let us assume for concreteness that $\rho_1 \geq 0$ (the case $\rho_1 < 0$ is considered in an analogous manner). Let us choose the number μ such that the spectrum of the operator $B_\mu = B_0 - \mu I$ is divided into two parts by the imaginary axis, the spectral points with the largest real parts being sufficiently close to the imaginary axis. Let us enclose the eigenvalues of the operator $B_0 - \mu I$ situated in the right half-plane by circles of sufficiently small radius δ; let ν be the distance of these circles from the imaginary axis. Then, it is obvious that the exponential-growth index for the solutions of Eq. (4.2) is

$$\varepsilon\rho_1 = \varepsilon(\mu + \nu + \delta).$$

When ε is sufficiently small, Eq. (4.4) will have indefinitely increasing solutions, so that the index $\rho = \rho(\varepsilon)$ of the exponential growth of the solutions of Eq. (4.1) is not less than $\varepsilon\mu$, i.e.,

$$\rho(\varepsilon) \geqslant \varepsilon\mu.$$

We therefore have

$$\rho(\varepsilon) - \varepsilon\rho_1 \geqslant -\varepsilon(\nu + \delta).$$

Let us now choose μ such that the whole of the spectrum of the operator $B_\mu = B_0 - \mu I$ is in the left half-plane, but the distance between the imaginary axis and the spectrum is sufficiently small. Let us again enclose the eigenvalues with the largest real parts by circles of sufficiently small radius δ and let ν_1 denote the distance of these circles from the imaginary axis. Then, we will obviously have

$$\varepsilon\rho_1 = -\varepsilon\mu \geqslant -\varepsilon(\nu_1 + \delta).$$

When ε is sufficiently small all solutions of Eq. (4.4) will asymptotically tend to zero, i.e., we have

$$\rho(\varepsilon) \leqslant \varepsilon\mu.$$

We therefore have

$$\rho(\varepsilon) - \varepsilon\rho_1 \leqslant \varepsilon(\nu_1 + \delta).$$

Thus, we see that

$$-\varepsilon(\nu + \delta) \leqslant \rho(\varepsilon) - \varepsilon\rho_1 \leqslant \varepsilon(\nu_1 + \delta).$$

Choosing, for example, $\nu = \nu_1 = \delta$ and making use of the inequalities of Lemmas 8 and 9, we obtain

$$|\rho(\varepsilon) - \varepsilon\rho_1| < C\varepsilon^{1+\frac{1}{4q-2}},$$

i.e., inequality (4.3).

If the elementary divisors of the eigenvalues of the operator B_0 are prime, then the estimate given above can be improved. Indeed, in this case we have $\|P_1 B_\mu\| \leqslant \nu C$, where ν is the distance between the contours being considered and the imaginary axis and C is a constant. Consequently, we have $\alpha_1 \geq 1/2\nu C$ and the inequalities of Lemma 9 become

$$\varepsilon < \varepsilon_0, \quad \delta < \delta_0, \quad \varepsilon C < \nu,$$

from which it follows that if we let δ tend to zero and choose $\nu = \varepsilon C$, then we can take $\delta = 1$ in inequality (4.3). Theorem 1 has been proved.

If the spectral points of the operator B_0 with the largest real parts are not isolated, but the resolvent $(B_0 - \lambda I)^{-1}$ of this operator satisfies the inequality

$$\|(B_0 - \lambda I)^{-1}\| < \frac{C}{\rho^q(\lambda)}, \tag{4.22}$$

where C is a constant and $\rho(\lambda)$ is the distance of the point λ from the spectrum of B_0, then we have the following weaker version of Theorem 1.

THEOREM 2. Let the resolvent of the operator B_0 satisfy conditions (4.22). Let ρ_1 and ρ_2 denote the largest and smallest real parts of the discrete spectrum of B_0. Then the following inequality holds for an arbitrary sufficiently small $\varepsilon > 0$:

$$\varepsilon\rho_2 - C\varepsilon^{1+\gamma} < \rho(\varepsilon) < \varepsilon\rho_1 + C\varepsilon^{1+\gamma}, \tag{4.23}$$

where C and γ are positive constants depending on the operator B_0. If we have $q = 1$ in Inequality (4.22), then we can take $\gamma = 1$ in inequalities (4.23).

The proof of Theorem 2 consists of the appropriate parts of the proof of Theorem 1 and is therefore omitted.

Remark. In the case when the whole spectrum of B_0 is situated on the imaginary axis, we have $\rho_1 = 0$. It then follows from (4.23) that the index of the exponential growth of solutions of Eq. (4.1) is a quantity of a higher order of smallness than ε.

§ 5. The Applicability of the Method of Averaging

If the operator function $J^{-1}H(t)$ satisfies the conditions of Lemma 5, then in view of this lemma there exists a change of variable

$$z = R(\varepsilon, t)x, \tag{5.1}$$

where $R(\varepsilon, t)$ is an operator function that is uniformly bounded with respect to t, with $-\infty < t < \infty$, for all sufficiently small ε. Equation (1.1) reduces to Eq. (4.1) with the help of this transformation. However, it is a fairly difficult matter to verify that the conditions of Lemma 5 are satisfied since it is necessary to investigate the asymptotic properties of the eigenvalues of the operator $iJ^{-1}H_0$. The aim of the present section is to show that transformation (5.1) exists when the operator $iJ^{-1}H_0$ possesses only the Property S_δ (see p. 30), but Eq. (1.1) is Hamiltonian and the lacunae of $iJ^{-1}H_0$ are sufficiently large. The rigorous assertion can be stated as the following theorem.

THEOREM 3. Let us assume that the operator $iJ^{-1}H_0$ for the Hamiltonian equation (1.1) possesses Property S_δ (see definition 2) and in addition

$$\delta > \max_k |(\omega, k)|,$$

where the maximum is taken over all integer vectors k corresponding to nonzero coefficients of $H^{(k)}$ in the summation (1.2). Then, for all sufficiently small ε there exists a function $R(\varepsilon, t)$ defined for all real t with the following properties:

1) The values of this function are linear operators bounded in space **H** whose inverses are also bounded;

2) The norms of the functions $R(t, \varepsilon)$ and $R^{-1}(t, \varepsilon)$ are uniformly bounded with respect to t, $-\infty < t < \infty$;

3) Eq. (1.1) reduces to Eq. (4.1) with the help of the transformation (5.1). The operator B_0 in Eq. (4.1) is finite-dimensional and acts in subspace P_2**H**, where $P_2 = \int_\Delta dE_\lambda$, E_λ being the spectral family [20] of the operator $iJ^{-1}H_0$ and Δ the region of the spectrum situated between the lacunas entering the definition of Property S_δ.

Proof. Let q_1 and q_2 (where $q_1 < 0$ and $q_2 > 0$) be the centers of lacunas whose widths are not less than δ, and let P_1, P_2, and P_3 be the orthoprojectors on the characteristic subspaces of the operator $iJ^{-1}H_0$ corresponding to regions of the spectrum that are situated to the left of q_1, between q_1 and q_2, and to the right of q_2, respectively. According to Lemma 6, the operator function

$$\Phi(t) = B(t) - \sum_{i=1}^{3} P_i B(t) P_i, \tag{5.2}$$

where

$$B(t) = e^{-J^{-1}H_0 t} J^{-1} H(t) e^{J^{-1}H_0 t} \tag{5.3}$$

does not have Fourier indices in a neighborhood of zero. According to Lemma 5, the operator function

$$\Psi(t) = \int_0^t \Phi(s)\, ds \tag{5.4}$$

is a bounded operator from **H** into **C**. Let us consider the change of variable

$$\tilde{z} = (I + \varepsilon\Psi(t))^{-1} e^{-J^{-1}H_0 t} x, \tag{5.5}$$

where the operator function $\Psi(t)$ is defined by formulas (5.2)-(5.4). If the parameter ε is sufficently small, then the operator $(I + \varepsilon\Psi(t))^{-1}$ is close to the unit operator uniformly with respect to t and, consequently, the change of variable (5.5) is correct for all t. Differentiating $\tilde{z}$ with respect to t and taking Eq. (1.1) into account, we obtain

$$\frac{d\tilde{z}}{dt} = -\varepsilon R_1(t, \varepsilon)\left[\frac{d\Psi(t)}{dt} - B(t)\right] R_1^{-1}(t, \varepsilon)\tilde{z}, \tag{5.6}$$

where for brevity we have used the abbreviation

$$R_1(t, \varepsilon) \equiv (I + \varepsilon \Psi(t))^{-1}.$$

Taking formulas (5.4), (5.3), and (5.2) into account, we can rewrite (5.6) as

$$\frac{d\tilde{z}}{dt} = \varepsilon R_1(t, \varepsilon) \sum_{l=1}^{3} P_l B(t) P_l R_1^{-1}(t, \varepsilon) \tilde{z}.$$

Finally, with the help of the identity

$$R_1(t, \varepsilon) = I - \varepsilon R_1(t, \varepsilon) \Psi(t),$$

valid for sufficiently small ε, we obtain the following equation for $\tilde{z}$:

$$\frac{d\tilde{z}}{dt} = \varepsilon \sum_{l=1}^{3} P_l B(t) P_l \tilde{z} + \varepsilon^2 B_1(t, \varepsilon) \tilde{z}, \tag{5.7}$$

where $B_1(t, \varepsilon)$ is an operator function uniformly bounded with respect to t for sufficiently small ε. Let us consider the equation

$$\frac{dz}{dt} = \varepsilon \sum_{l=1}^{3} P_l B(t) P_l z \tag{5.8}$$

in greater detail. Inasmuch as the operators $P_i B(t) P_i$ with different indices act in mutually orthogonal subspaces, the operator solution Z(t), with Z(0) = I, of Eq. (5.8) can be written as

$$Z(t) = Z_1(t) Z_2(t) Z_3(t), \tag{5.9}$$

where the operator functions $Z_i(t)$ are defined by the equations

$$\frac{dZ_l}{dt} = \varepsilon P_l B(t) P_l Z_l, \quad Z_l(0) = I, \quad l = 1, 2, 3. \tag{5.10}$$

The following properties of the functions under consideration are obvious:

$$\begin{gathered} Z_l(t) Z_j(t) = Z_j(t) Z_l(t), \\ Z_l(t) P_j B(t) P_j = P_j B(t) P_j Z_l(t), \quad l \neq j. \end{gathered} \tag{5.11}$$

Up to this point, we have not made use of the fact that the equation is Hamiltonian. Let us show that if Eq. (1.1) is Hamiltonian, then the operator functions $Z_1(t)$ and $Z_3(t)$ are unitary operators. Let us consider for concreteness the equation with index three. In view of the fact that the operators J and H_0 commute, it follows from formula (5.3) that iJB(t) is a self-adjoint operator. Therefore, this means that Eqs. (5.10) are Hamiltonian. The operator $iJ^{-1}H_0$ is positive definite on the subspace $P_3\mathbf{H}$ since the whole of its spectrum is situated to the right of $q_2 > 0$. Since H_0 is a positive operator in **H**, the operator iJ^{-1} is positive definite on the subspace $P_3\mathbf{H}$. Denoting the positive root of the operator $\frac{1}{i} JP_3$ by Q, we can use the transformation $\tilde{\tilde{z}}_3 = Q^{1/2} z_3$ to arrive at the Schroedinger equation

$$i \frac{d\tilde{\tilde{z}}_3}{dt} = \varepsilon P_3 \tilde{\tilde{B}}(t) P_3 \tilde{\tilde{z}}_3,$$

where B(t) is a self-adjoint operator that is bounded for all t. It is well-known that the operator solution of the Schroedinger equation is a unitary operator for all t. Because Q commutes with H_0, the operator $Z_3(t)$ is also unitary. We prove that $Z_1(t)$ is a unitary operator in an analogous manner.

Let us now perform the change of variable

$$\tilde{\tilde{z}} = Z_1^{-1}(t) Z_2^{-1}(t) \tilde{z} \tag{5.12}$$

in Eq. (5.7); here, $Z_1(t)$ and $Z_2(t)$ are the operator solutions of Eq. (5.10) for the indices i = 1, 3. Differentiating z with respect to t and taking (5.7) and (5.11) into account, we obtain

$$\frac{d\tilde{\tilde{z}}}{dt} = [\varepsilon P_2 B(t) P_2 + \varepsilon^2 B_2(t, \varepsilon)] \tilde{\tilde{z}}, \tag{5.13}$$

where

$$B_2(t, \varepsilon) = [Z_1(t) Z_3(t)]^{-1} B_1(t, \varepsilon) Z_1(t) Z_3(t) \tag{5.14}$$

is an operator function that is uniformly bounded with respect to t for a sufficiently small ε. The operator function $P_2B(t)P_2$ has only a finite number of Fourier indices because $J^{-1}H_0$ is a completely continuous operator and, consequently, P_2 is a projector on a finite-dimensional space. Therefore, zero cannot be a point of condensation of the Fourier indices of this function. Let B_0 denote the average value of the function $P_2B(t)P_2$; then

$$\tilde{\Psi}(t) = \int_0^t P_2 B(s) P_2 \, ds - B_0 t$$

is an operator function that acts in a bounded manner from **H** into **C** and the change of variable

$$z = (I - \varepsilon\tilde{\Psi}(t))^{-1} \tilde{\tilde{z}}$$

will reduce Eq. (5.13) to Eq. (4.2). Moreover, it is obvious that the operator B_0 acts in the subspace P_2**H**. The theorem has been proved.

Thus, when the conditions of Theorem 3 are satisfied, the conditions of Theorem 1 are automatically satisfied and, consequently, the infinite-dimensional Eq. (1.1) is equivalent in the approximation under consideration to a system with a finite number of degrees of freedom.

It follows from the proof of Theorem 3 that the main part of the theorem remains valid for a quasi-Hamiltonian equation if more stringent conditions are imposed on the operator $iJ^{-1}H_0$, namely, we must require that the spectral points sufficiently far removed from the origin be separated from one another by a distance not less than δ. In this case, as can be easily seen, the Fourier indices of the operator functions $P_1B(t)P_1$ and $P_3B(t)P_3$ cannot condense to zero. Consequently, the function $\sum_{i=1}^{3} P_i B(t) P_i$ satisfies all the conditions of Lemma 5. Moreover, the operator B_0 in the averaged equation commutes with the orthoprojectors P_i, i = 1, 2, 3. Finally, it should be noted that many problems in the mechanics of structures lead to operators $iJ^{-1}H_0$ possessing Property S in a strong form: the distance between neighboring eigenvalues increases without limit as their absolute values increases.

PART II

THE CONSTRUCTION OF THE DOMAINS OF DYNAMIC INSTABILITY FOR QUASI-STATIONARY EQUATIONS WITH ALMOST-PERIODIC COEFFICIENTS

§ 6. Formulation of the Problem of Parametric Resonance

Many problems in physics and mechanics lead to the study of the behavior as $t \to \infty$ of solutions of Hamiltonian and quasi-Hamiltonian equations of the form

$$J \frac{dx}{dt} = [H_0 + \varepsilon H(t, \omega)]\, x, \tag{6.1}$$

whose coefficients satisfy the conditions formulated in Section 1. Equation (6.1) is said to be *stable* if all of its solutions are bounded for $t \to \infty$, otherwise it is said to be *unstable*.

As has already been pointed out in Section 1, the operator solution of the unperturbed equation (i.e., Eq. (6.1) with $\varepsilon = 0$) is a unitary operator for all t and, consequently, the unperturbed equation is stable. In the presence of a perturbation with an arbitrarily small amplitude, Eq. (6.1) may no longer be stable. The simplest cases of reducible systems* show that this situation is the rule rather than an exception in the case of systems with an infinite number of degrees of freedom. More accurately, for a very wide class of quasi-Hamiltonian equations, the perturbed equation can be made unstable by an arbitrarily small change in the frequency vector ω [9, 10, 12]. Therefore, in the case of the systems being considered, it is not the unboundedness or boundedness of the oscillations that is of greatest interest, but the rate of growth of the oscillations as $t \to \infty$. This rate is governed by the exponential-growth index $\rho(\varepsilon, \omega)$ defined by formula (1.7) for the solutions of Eq. (6.1). It is important for applications to be able to derive formulas that represent effective procedures for identifying domains in the space of the parameters $\{\varepsilon, \omega\}$ where $\rho(\varepsilon, \omega)$ has its maximum values. Let us introduce the following definition as an aid to a more accurate formulation of such problems.

Definition 3. a) *The frequency vector $\omega = \omega_0$ is said to be a resonance vector if there exists a vector ω_1 such that in the $(n + 1)$-dimensional parameter space $\{\varepsilon, \omega\}$†, the inequality*

$$\lim_{\varepsilon \to 0} \varepsilon^{-1} \rho(\varepsilon, \omega_0 + \varepsilon\omega_1) > 0 \tag{6.2}$$

holds on the ray $\{\varepsilon, \omega_0 + \varepsilon\omega_1\}$.‡

b) *The set of all rays of the above type for small values of ε will be called the set of essential instability of the solutions of Eq. (6.1) adjoining the point $\{0, \omega_0\}$.*

---

* That is, equations whose operator coefficients in some basis of space H have the form of quasi-diagonal matrices with blockes of finite size.

† n is the number of dimensions of the space of the frequency vectors ω.

‡ Thus, if ω_0 is a resonance vector, then the solutions of the perturbed equation increase exponentially along a ray beginning at the point $\{0, \omega_0\}$, the exponent being proportional to ε to within $o(\varepsilon)$. On the other hand, if ω_0 is not a resonance vector, then the exponential-growth index along any ray beginning at $\{0, \omega_0\}$ is either negative or is a small quantity of order $o(\varepsilon)$.

c) The set in parameter space $\{\varepsilon, \omega\}$ defined by the inequality $\rho(\varepsilon, \omega) > \alpha$ for a given number α is called the set of α-exponential growth of the solutions of Eq. (6.1).

The problem consists in the determination of the resonance frequency vectors for Eq. (6.1), the construction of the corresponding sets of essential instability, and the calculation of the perturbation amplitude required to produce α-exponential growth along a given ray. The ability to derive the sets of essential instability and to construct the corresponding sets of α-exponential growth of solutions allow us to investigate many concrete systems in the theory of the dynamic stability of elastic systems.

§ 7. The Nonresonance Case

In those cases when the main term in the averaged Eq. (4.1) – the operator B_0 – is a null operator, the function $\rho(\varepsilon, \omega)$ is of the second order of smallness in ε and, consequently, the solutions of the perturbed equation cannot grow "too fast" as $t \to \infty$. The present section is devoted to the investigation of such "nonresonance" cases.

In the following, let us agree to enumerate the eigenvalues λ_j of the operator $iJ^{-1}H_0$,

$$iJ^{-1}H_0 a_j = \lambda_j a_j, \quad j = \pm 1, \pm 2, \ldots, \tag{7.1}$$

in order of increasing $|\lambda_j|$ so that we have $\lambda_j \operatorname{sign} j > 0$. Since the operators J^{-1} and B_0 commute, we can assume without loss of generality that

$$iJ^{-1}a_j = \operatorname{sign} j a_j\text{*}. \tag{7.2}$$

THEOREM 4. Let us assume that for the frequency vector $\omega = \omega_0$ a relation of the form

$$(\omega_0, k) + \lambda_l - \lambda_m = 0 \tag{7.3}$$

does not hold for any eigenvalues λ_l, λ_m of the operator $iJ^{-1}H_0$ and for any integer vectors k that correspond to nonzero Fourier coefficients $H^{(k)}$ of the function $H(t, \omega)$. Let us assume, in addition, that either the conditions of Theorem 3 hold or that the Fourier indices of the function

$$B(t) = e^{-J^{-1}H_0 t} J^{-1} H(t, \omega) e^{J^{-1}H_0 t} \tag{7.4}$$

do not have a condensation point at zero.

Then, 1) the frequency ω_0 cannot be a resonance vector,

2) an arbitrary vector ω_1 satisfies the inequality

$$|\rho(\varepsilon, \omega_0 + \varepsilon\omega_1)| < C\varepsilon^2, \tag{7.5}$$

where the constant C can be chosen to be uniform with respect to changes in the vector ω_1 in every bounded set.

If relation (7.3) holds for some eigenvalues λ_l, λ_m and an integer vector k different from the zero vector, then the class of all quasi-

---

*If conditions (7.2) are not satisfied, then a nonsingular bounded transformation $y = |J|^{1/2}x$, where $|J| = (-J^2)^{1/2}$, allows us transform Eq. (6.1) into a quasi-Hamiltonian equation in which the new operator J will possess the required property. This transformation does not affect the investigation of the stability of solutions.

Hamiltonian equations of the form (6.1) will contain an equation for which ω_0 will be a resonance vector.

Proof. In view of Definition 3, the first assertion of the theorem is a consequence of the second. We therefore proceed to prove the second assertion.

Because of the conditions of the theorem, relation (7.3) cannot hold for any vector ω belonging to a sufficiently small neighborhood of the vector ω_0. Let us now make use of the results of Section 3. If $\omega = \omega_0 + \varepsilon\omega_1$, then the operator function $\Psi(t)$ defined by formula (3.6) will depend on ε, but, because relation (7.3) does not hold for the vector ω, it will be uniformly bounded with respect to t, $-\infty < t < \infty$, for all sufficiently small ε. Therefore, the term involving ε^2 in Eq. (4.1) will be uniformly bounded with respect to t and ω_1 for all $\omega = \omega_0 + \varepsilon\omega_1$, provided that ε is sufficiently small and ω_1 belongs to any bounded set. Let us now consider the principal term in Eq. (4.1) – the operator B_0. It is obvious that the collection of normalized eigenfunctions $\{a_j\}$ of the operator $iJ^{-1}H_0$ represents an orthonormal basis in space **H**. In the basis $\{a_j\}$, the operator B_0 – the average value of the operator function B(t) – has the form of a matrix whose elements are the numbers

$$[B_0]_{lm} = i \operatorname{sign} m \sum_k (H^{(k)}a_l, a_m)\, \delta[(\omega, k) + \lambda_l - \lambda_m], \tag{7.6}$$

where

$$\delta(s) = \begin{cases} 0, & s \neq 0, \\ 1, & s = 0. \end{cases}$$

Since relation (7.3) does not hold for all ω_1 sufficiently close to ω_0, it follows from formula (7.6) that for sufficiently small ε the operator B_0 is a null operator. It is obvious that Theorem 1 remains valid and leads to inequality (7.5) in which the constant C can be chosen independent of the vector ω_1 when the latter varies within each bounded set.

The last assertion of the theorem is obvious since formulas (7.6) show that if relation (7.3) is satisfied for some λ_l, λ_m, and k, then we can choose operators $H^{(k)}$ such that B_0 has a spectrum in the right half-plane. The theorem has been proved.

Theorem 4 asserts, in particular, that in the space $\{\omega\}$ the resonance vectors can only be situated on hyperplanes of the form (7.3) and, consequently, the set of resonance vectors is of measure zero. However, it is just this set that is of especial interest in practice.

If we are considering a subclass of the class of quasi-Hamiltonian equations, then the condition that a relation of the type of (7.3) holds may not be a necessary condition for the vector ω_0 to be a resonance vector. For example, in the theory of Hamiltonian equations with periodic coefficients (in this case the frequency vectors are scalar quantities) there exists the well-known theorem of M. G. Krein [22] according to which resonance frequencies ω_0 can only satisfy relation (7.3) when the numbers λ_l and λ_m are of opposite signs. It is found that the analogous situation also holds in the case of Hamiltonian equations with almost-periodic coefficients.

THEOREM 5. Let us assume that the conditions of Theorem 3 in the case of Eq. (7.1) hold for the frequency vector $\omega = \omega_0$. Let Λ denote the set of eigenvalues of the operator $iJ^{-1}H_0$ belonging to the central part of its spectrum.* If

* That is, Λ is the set of eigenvalues situated between the lacunas mentioned in Theorem 3. This set is finite because of the assumptions made above concerning Eq. (6.1).

$$(H^{(k)}a_l,\ a_m)=0 \tag{7.7}$$

holds for any numbers λ_l, λ_m belonging to Λ, satisfying condition (7.3), and being of opposite signs, then assertions 1) and 2) of Theorem 4 are valid.

If relation (7.3) holds for numbers λ_l and λ_m of opposite signs belonging to Λ and for an integer vector k, then the class of Hamiltonian Eqs. (6.1) will contain an equation for which ω_0 will be a resonance vector.

Proof. Let us return to the proof of Theorem 3. Assuming that $\omega = \omega_0 + \varepsilon\omega_1$, where ω_1 is an arbitrary fixed vector, we can again assert that function (5.2) does not possess Fourier indices in a neighborhood of zero for all sufficiently small ε, and, consequently, function (5.4) for these ε is a bounded operator from **H** into **C**. Repeating the arguments of Theorem 3, we arrive at Eq. (5.13) with a function of $B(t, \varepsilon)$ uniformly bounded with respect to t if ε is sufficiently small. The function $P_2B(t)P_2$ may have zero Fourier indices when $\omega_1 = 0$. When $\omega_1 \neq 0$, these indices being continuous functions of ε may become different from zero and the method of averaging cannot be carried out for Eq. (5.13). We will therefore proceed as follows. Let us take a positive number δ such that for sufficiently small ε the function $P_2B(t)P_2$ has in the δ-neighborhood of zero only those Fourier indices that become zero when $\omega_1 = 0$. Let us represent $P_2B(t)P_2$ as

$$P_2B(t)P_2 = D_1(t,\ \varepsilon) + D_2(t,\ \varepsilon), \tag{7.8}$$

where for all sufficiently small ε the function $D_1(t, \varepsilon)$ has Fourier indices only outside the δ-neighborhood of zero and $D_2(t, \varepsilon)$ has Fourier indices only inside the δ-neighborhood of zero. The function

$$\tilde{\Psi}(t) = \int_0^t D_1(t,\ \varepsilon)\,dt \tag{7.9}$$

will then be uniformly bounded with respect to t, $-\infty < t < \infty$ for these values of ε. Let us perform in Eq. (5.13) the substitution

$$y = (I + \varepsilon\tilde{\Psi}(t))^{-1}\tilde{\tilde{z}}; \tag{7.10}$$

y will satisfy the equation

$$\frac{dy}{dt} = \varepsilon D_2(t,\ \varepsilon)\,y + \varepsilon^2 D_3(t,\ \varepsilon)\,y, \tag{7.11}$$

where for sufficiently small values of ε the functions $D_2(t, \varepsilon)$ and $D_3(t, \varepsilon)$ are uniformly bounded with respect to t and the function $D_2(t, \varepsilon)$ has Fourier indices only in the δ-neighborhood of zero. It follows from formulas (5.3) and (7.8) that the function $JD_2(t, \varepsilon)$ is self-adjoint. Up till now we have not used the condition of the theorem that only numbers λ_l and λ_m with special properties can appear in relations (7.3). Making use of these properties, we will show that the operator function $D_2(t, \varepsilon)$ is anti-Hermitian. Indeed, let us consider this function in the basis $\{a_j\}$ formed by the eigenfunctions of the operator $iJ^{-1}H_0$ in the subspace P_2**H**. The corresponding matrix elements are

$$(D_2(t,\ \varepsilon)\,a_l,\ a_m) = \begin{cases} i\sum_k e^{i[\lambda_l-\lambda_m+(\omega,\ k)]t}\operatorname{sign} m\,(H^{(k)}P_2a_l,\ P_2a_m), \\ \qquad\qquad \text{when } |\lambda_l-\lambda_m+(\omega_0,\ k)|=0, \\ 0, \qquad\quad \text{when } |\lambda_l-\lambda_m+(\omega_0,\ k)|\neq 0, \end{cases} \tag{7.12}$$

$$\omega = \omega_0 + \varepsilon\omega_1.$$

But, by definition, for the numbers λ_l and λ_m satisfying condition (7.3) we have either $(D_2(t, \varepsilon) a_l, a_m) = 0$ or sign l = sign m. Taking into account that because the function H(t, ω) is self-adjoint, we have

$$[H^{(k)}]^* = H^{(-k)}$$

and we find from formula (7.12) that

$$(D_2(t, \varepsilon) a_l, a_m) = -\overline{(D_2(t, \varepsilon) a_m, a_l)},$$

from which it follows that the function $D_2(t, \varepsilon)$ is anti-Hermitian. In this case, however, the operator solution $\tilde{Y}(t)$ of the equation

$$\frac{d\tilde{y}}{dt} = \varepsilon D_2(t, \varepsilon) \tilde{y}$$

is a unitary operator. The change of variable

$$z = \tilde{Y}^{-1} y$$

reduces Eq. (7.11) to the equation

$$\frac{dz}{dt} = \varepsilon^2 \tilde{Y}(t, \varepsilon) D_3(t, \varepsilon) \tilde{Y}^{-1}(t, \varepsilon) z \tag{7.13}$$

with an operator function that is uniformly bounded with respect to t for all sufficiently small values of ε. In view of Theorem 1, the exponential-growth index $\rho(\varepsilon, \omega_0 + \varepsilon\omega_1)$ for the solutions of Eq. (7.11) satisfies the bound

$$\rho(\varepsilon, \omega_0 + \varepsilon\omega_1) \leqslant C\varepsilon^2 \tag{7.14}$$

with a constant C that, as was already noted above, can be chosen from a selected set to be uniform with respect to ω_1. Inasmuch as the substitutions reducing the original Eq. (6.1) to Eq. (7.13) were uniformly bounded with respect to t, the exponential growth indices for the solutions of these equations are equal. It follows from inequality (7.14) that the vector ω_0 is not a resonance vector. The first part of the theorem has been proved. The second part of the theorem follows directly from the formulas of the first approximation that will be derived below in Section 9.

Remark. Theorem 5 asserts that the frequency vectors ω_0 that are resonance vectors for the Hamiltonian equations can be situated in $\{\omega\}$ space only on the hyperplanes

$$(\omega_0, k) = \lambda_m - \lambda_l, \quad \lambda_m\lambda_l < 0, \tag{7.15}$$

the set of which is much more sparse than the corresponding set of hyerplanes for quasi-Hamiltonian equations.

In the following we will only be interested in cases of resonance and, in particular, in the problem of the determination of all those vectors ω_0 that satisfy inequality (6.2).

In those cases when the operator B_0 in the averaged equation is a null operator, we can try to perform further averaging, extracting terms with ε^2 independent of t, etc. This will allow us to obtain more accurate expressions for the function $\rho(\varepsilon, \omega)$ for small ε. Transformations of this type will be examined in the third part of the present article.

§ 8. Resonance Perturbation Frequencies. The Concept of Degeneracy

Making use of the representation of the operator B_0 in the basis formed by the eigenfunctions a_j of the operator $iJ^{-1}H_0$ [see formulas (7.6)], we can obtain an effective formula for the quantity $\varepsilon\rho_1$ which, in view of Theorem 1, represents the principal part of the function $\rho(\varepsilon, \omega)$ of interest for small ε. In this connection, however, it is not difficult to see that the role of the "principal" part played by $\varepsilon\rho_1$ is very "unstable" with respect to small changes in the vector ω. More accurately, this means the following. Let us assume that ω_0 is a resonance vector. Then, in view of Theorem 4, relation (7.3) holds for some k, l, and m. It follows from formula (7.3) that by making arbitrarily small changes in the vector ω_0 we can make the quantity

$$(\omega, k) + \lambda_l - \lambda_m \tag{8.1}$$

be as close as we like to zero, but different from zero. But in this case the quantity (8.1) is the denominator of one of the terms of the Fourier series of the function $\Psi(t)$ realizing the transformation of Eq. (1.1) into Eq. (4.1). Therefore, for any fixed value of ε, an arbitrarily small change in the vector ω_0 can lead to a situation in which the term $\varepsilon B_0 x$ in Eq. (4.1) is no longer the "principal" term. Such a nonuniform dependence of the coefficients of the averaged Eq. (4.1) on ω complicates the investigation of the dependence of $\rho(\varepsilon, \omega)$ on its second argument in the neighborhood of the resonance frequency vectors.

It can be seen from expressions (8.1) that if we assume that the eigenvalues λ_l, λ_m of the operator $iJ^{-1}H_0$ also change with changes in ω, then, at least in some cases, we can hope that the quantity (8.1) will remain zero as ω changes and, consequently, that small numerators will no longer appear in the numerators of the Fourier series of the operator function $\Psi(t)$. The realization of this idea leads to a transformation of Eq. (4.1) which in a definite sense allows us to weaken the dependence of the coefficients of the averaged equation on ω. A procedure of this type allows us to overcome the above mentioned difficulty to the end only in the so-called nondegenerate case which will be explained below. Before we proceed to a formal definition of the concept of degeneracy, let us consider a transformation which leads to this concept in a natural manner and which is an insignificant modification of the method of averaging presented in Section 3. At present, let us only note that the requirement of nondegeneracy is a condition imposed on the unperturbed frequency vector ω_0 and that this condition holds for almost all resonance vectors ω_0.*

Let Q be a self-adjoint operator, the inverse of some operator Q^{-1}, and let Q be continuous and commute with the operator $J^{-1}H_0$

$$Q^{-1}J^{-1}H_0 = J^{-1}H_0Q^{-1}. \tag{8.2}$$

Let us consider the change of variable

$$z = (I + \varepsilon\Psi(t))^{-1}e^{iQt}x, \tag{8.3}$$

where

$$\Psi(t) = \int_0^t e^{iQs}J^{-1}H(s, \omega)e^{-iQs}ds - Bt, \tag{8.4}$$

$$B = \lim_{t\to\infty}\frac{1}{2t}\int_{-t}^{t} e^{iQs}J^{-1}H(s, \omega)e^{-iQs}ds. \tag{8.5}$$

The limit in (8.5) is to be understood in the strong sense; it exists in view of Lemma 4. We will assume that the operator Q satisfies the conditions of Lemma 5, so that the operator func-

* Strictly speaking, the last assertion is only valid when the set in frequency $\{\omega\}$ of interest to us does not contain vectors orthogonal to the integer vectors k appearing in the summation (1.2).

tion $\Psi(t)$ is uniformly bounded with respect to t, and consequently, the transform (7.3) is significant for all t provided that the positive parameter ε is sufficiently small. Assuming that x is a solution of Eq. (6.1), it is not difficult to show that z satisfies the equation

$$\frac{dz}{dt} = [J^{-1}H_0 + iQ]\, z + \varepsilon Bz + \varepsilon\, [\Psi(t)(J^{-1}H_0 + iQ) - (J^{-1}H_0 + iQ)\Psi(t)]\, z + \varepsilon^2 \Phi(\varepsilon,\ t)\, z, \tag{8.6}$$

where $\Phi(\varepsilon, t)$ is an operator function that is uniformly bounded with respect to t for all sufficiently small values of ε. When $iQ = J^{-1}H_0$, Eq. (8.6) turns into Eq. (4.1). In the following, we will be interested in the case when

$$J^{-1}H_0 + iQ = \varepsilon D, \tag{8.7}$$

where D is a finite-dimensional operator, the choice of which will be discussed later. Thus, instead of Eq. (4.1), we arrive at the equation

$$\frac{dz}{dt} = \varepsilon\,(B + D)\, z + \varepsilon^2 B\,(t,\ \varepsilon)\, z \tag{8.8}$$

with an operator function $B(t, \varepsilon)$ uniformly bounded with respect to t, $-\infty < t < \infty$, for all sufficiently small values of ε. It is obvious that the exponential-growth indices for the solutions of Eqs. (1.1), (4.1), and (8.1) are the same for all these values of ε. Let $\hat{\lambda}_j$ denote the eigenvalues of the operator Q, i.e., $Qa_j = \hat{\lambda}_j a_j$, and then we have

$$\hat{\lambda}_j = \lambda_j + \varepsilon\mu_j, \tag{8.9}$$

where $i\mu_j$ are the eigenvalues of the operator D. The operator B defined by formula (8.5) has the form of a matrix $\{B_{lm}\}$ in the basis formed by the eigenfunctions a_j of the operator $iJ^{-1}H_0$, the matrix elements B_{lm} being the numbers

$$B_{lm} = i \sum_k (H^{(k)}a_l,\ a_m)\, \mathrm{sign}\, m\delta\, [(\omega,\ k) + \hat{\lambda}_l - \hat{\lambda}_m], \tag{8.10}$$

where $\delta(s)$ is defined by formulas (7.5). Let us now assume that we have

$$\omega = \omega_0 + \varepsilon\omega_1, \tag{8.11}$$

where ω_0 is a fixed vector and ω_1 an arbitrary vector that characterizes the perturbation of the frequency vector. In this case, the expression forming the argument of the function $\delta(S)$ in formula (8.10) can be written as

$$(\omega_0,\ k) + \lambda_l - \lambda_m + \varepsilon\, [(\omega_1,\ k) + \mu_l - \mu_m]. \tag{8.12}$$

If expression (8.12) is equal to zero and we have $(H^{(k)}a_l, a_m) \neq 0$, then the corresponding element B_{lm} will in general be nonzero. If expression (8.12) is equal to zero for all ε, then we have

$$(\omega_0,\ k) + \lambda_l - \lambda_m = 0, \tag{8.13}$$

$$(\omega_1,\ k) + \mu_l - \mu_m = 0. \tag{8.14}$$

Relation (8.13) was discussed above and it is satisfied by the resonance frequency vector ω_0. The case which was referred to earlier as the nondegenerate case is characterized by the property that, given the vector ω_1, it is possible to select numbers μ_j such that for any vectors k satisfying relations of the type of (8.13), relations of the type of (8.14) would also hold. Indeed, the impossibility of satisfying all relations of the type of (8.14) by an appropriate choice of the μ_j defines degeneracy; in this case, the transformation performed above does not reach its goal and for small ε we obtain small denominators in the Fourier series of the function $\Psi(t)$

for the frequencies given by (8.11). Let us now proceed to a more detailed discussion of the notions associated with degeneracy.

Let K denote the set of all vectors k with integral components for which the corresponding Fourier coefficients $H^{(k)}$ in the sum (1.2) are different from null operators. The set K is finite because of the asssumptions made above.

Let ω_0 be a resonance frequency vector. Then, a vector k, with $k \in K$, satisfying relation (8.13) exists for some λ_l and λ_m. There may be several such relations for a fixed vector ω_0 corresponding to various $k \in K$ and various eigenvalues of the operator $iJ^{-1}H_0$. Let us denote by $\tilde{K}$, with $\tilde{K} = \tilde{K}(\omega_0)$, the set of all vectors $k \in K$, each of which satisfies a relation of the type of (8.13) (of course, with its own λ_l and λ_m). Let us number the vectors $k \in \tilde{K}$ in some manner. To each vector $k_l \in \tilde{K}$, where $l = 1, 2, \dots, N$, there may correspond a number N_l of distinct pairs $\lambda_{j_l} - \lambda_{i_l}$ satisfying a relation of the type of (8.13). If we write down all of these relations, we obtain the system

$$(\omega_0, k_l) = \lambda_{i_m^{(l)}} - \lambda_{j_m^{(l)}}, \quad \begin{array}{l} m = 1, 2, \dots, N_l, \\ l = 1, 2, \dots, N. \end{array} \tag{8.15}$$

Here, for each vector $\overline{k}_l$ we have written down as many relations as there are distinct pairs $\lambda_{i(l)} - \lambda_{j(l)}$ corresponding to it. System (8.15) thus consists of $\sum_{l=1}^{N} N_l$ relations. The number N_l may be infinite.)

Among the numbers λ_j appearing on the right-hand sides of relations (8.15) there may be numbers with identical indices.* Using the elimination method, we can transform system (8.15) into a form in which the right-hand sides do not contain numbers λ_j with identical indices. The transformed system can be written as

$$\left(\omega_0, \sum_{s=1}^{l} m_s^{(p)} k_s\right) = \sum_{m=1}^{N} \sum_{s=1}^{N_m} n_{sm}^{(p)} \lambda_{i_m^{(s)}}, \quad p = 1, 2, \dots, M, \tag{8.16}$$

where $m_s^{(p)}$, $n_{sm}^{(p)}$ and M are integers.

The quantity $\sum_{l=1}^{N} N_l - M$ characterizes the number of eliminations performed and, consequently, we have $M \leqslant \sum_{l=1}^{N} N_l$. In system (8.16), numbers λ_j with identical indices cannot have coefficients $n_{sm}^{(p)}$ that are simultaneously different from zero. We are now in a position to give the exact definition of the notion of degeneracy mentioned above.

Definition 4. If in system (8.16) obtained from system (8.15) by the elimination of numbers λ_j with identical indices, for any p such that

$$n_{sm}^{(p)} = 0, \; s = 1, 2, \dots, N_m; \; m = 1, 2, \dots, N, \tag{8.17}$$

the following relation holds simultaneously:

$$\sum_{s=1}^{l} m_s^{(p)} k_s = 0, \tag{8.18}$$

* It should be emphasized that we are interested in identical indices, and not identical numbers λ_j.

then we will say that the frequency vector ω_0 is nondegenerate. If for some p condition (8.18) does not follow from conditions (8.17), then the vector ω_0 will be said to be degenerate and, in this case, the number of nonzero vectors $\sum_{s=1}^{l} m_s^{(p)} k_s$ will be called the rank of the degeneracy of the vector ω_0.

Let us consider a few examples to illustrate the concept of the rank of degeneracy.

Example 1. Let system (8.15) be of the form

$$\begin{aligned}(\omega_0,\ k_1) &= \lambda_1 - \lambda_2,\\ (\omega_0,\ k_2) &= \lambda_1 - \lambda_2,\\ (\omega_0,\ k_3) &= \lambda_1 - \lambda_2,\\ (\omega_0,\ k_4) &= \lambda_1 - \lambda_2,\end{aligned}$$

where λ_1 and λ_2 are numbers, with $\lambda_1 \neq \lambda_2$, and k_i, with i = 1, 2, 3, 4, are distinct integer vectors. Eliminating λ_1 by means of the first equation of this system, written as $\lambda_1 = \lambda_2 + (\omega_0, k_1)$, we arrive at a system of the form of (8.16), namely,

$$\begin{aligned}(\omega_0,\ k_1 - k_2) &= 0,\\ (\omega_0,\ k_1 - k_3) &= 0,\\ (\omega_0,\ k_1 - k_4) &= 0.\end{aligned}$$

According to Definition 4, the vector ω_0 is degenerate and the rank of its degeneracy is equal to the number of vectors $k_1 - k_2$, $k_1 - k_3$, $k_1 - k_4$ that are different from the zero vector.

Example 2. Let system (8.15) be of the form

$$\begin{aligned}(\omega_0,\ k_1) &= \lambda_1 - \lambda_2,\\ (\omega_0,\ k_2) &= \lambda_2 - \lambda_1,\\ (\omega_0,\ k_3) &= \lambda_1 - \lambda_2,\\ (\omega_0,\ k_4) &= \lambda_2 - \lambda_1,\end{aligned}$$

where $k_1 = -k_2$, $k_3 = -k_4$, and $\lambda_1 \neq \lambda_2$. Eliminating λ_1, we obtain a system of the form of (8.16), namely,

$$\begin{aligned}(\omega_0,\ k_1 + k_2) &= 0,\\ (\omega_0,\ k_3 - k_1) &= 0,\\ (\omega_0,\ k_4 + k_1) &= 0.\end{aligned}$$

Since we have $k_1 = -k_2$, the first relation in the system we have obtained holds for all ω. If the vector $k_3 - k_1$ is different from the zero vector, then the vector ω_0 is degenerate and the rank of its degeneracy is equal to two.

The nondegenerate case can be characterized in terms of the numbers μ_j in the following manner.

THEOREM 6. Let us consider the system

$$(\omega_1,\ k_l) = \mu_{i_m^{(l)}} - \mu_{j_m^{(l)}}, \qquad \begin{matrix} m = 1,\ 2,\ \ldots,\ N_l, \\ l = 1,\ 2,\ \ldots,\ N, \end{matrix} \tag{8.19}$$

obtained from (8.15) by the replacement of the vector ω_0 by ω_1 and the numbers λ_j by μ_j. Let the frequency vector ω_0 be degenerate and let the rank of its degeneracy be equal to r. Then, the following assertions are valid:

1) There exists a vector ω_1 such that whatever the choice of the numbers μ_j at least r relations in system (8.19) will not hold.

2) For every vector ω_1 we can make a choice of the numbers μ_j in such a manner that all equations of system (8.19) are satisfied with the possible exception of some r equations.

Proof. Let the vector ω_0 be degenerate and let its rank of degeneracy be equal to r. Then, there exists a method for the step-by-step elimination of $\sum_{i=1}^{N} N_i - M$ numbers μ_j such that the remaining M distinct numbers must satisfy the system

$$\left(\omega_1, \sum_{s=1}^{N} m_s^{(p)} k_s\right) = \sum_{m=1}^{N} \sum_{s=1}^{N_m} n_{sm}^{(p)} \mu_{i_m^{(s)}}, \quad p = 1, 2, \ldots, M. \tag{8.20}$$

In addition, among the equations of system (8.20) there exists r equations whose right-hand sides are equal to zero identically with respect to μ_j, whereas the corresponding vectors $\sum_{s=1}^{N} m_s^{(p)} k_s$ are nonzero. It is obvious that we can select the vector ω_1 in such a manner that the right hand-sides of these r equations are different from zero. This means that whatever the choice of the numbers μ_j, there will be at least r relations in this system that will not hold for the given choice of the vector ω_1. The transformation of system (8.20) into system (8.19) is performed with the help of $\left(\sum_{i=1}^{N} N_i - M\right)$ elimination equations by means of multiplication by integers and the addition of the relations contained in system (8.20) and the elimination of equations. The number of inequalities in system (8.20) cannot decrease as the result of this procedure. Consequently, at least r equations of system (8.19) will not hold with the given choice of the vector ω_1. The first assertion of the theorem has been proved. In order to prove the second assertion, let us note that since system (8.20) only contains distinct numbers μ, for a given vector ω_1 we can arrange a selection of the numbers μ such that all relations (8.20) will hold with the possible exception of r relations. As before, the transition to system (8.19) with the help of $\left(\sum_{i=1}^{N} N_i - M\right)$ elimination equations cannot change the number of relations in system (8.20) that are not satisfied. This completes the proof of the theorem.

Remark. It follows from the assertions just proved that the concepts of degeneracy and the rank of degeneracy are independent of the method used to eliminate identical numbers μ_j in system (8.19).

Corollary to Theorem 4. In order for the vector ω_0 to be nondegenerate, it is necessary and sufficient that for every vector ω_1 the system (8.19) obtained from system (8.15) by the replacement of ω_0 by ω_1 and the numbers λ_j by μ_j be soluble for the numbers μ_j.

§ 9. The Nondegenerate Case

Let us assume that the resonance vector ω_0 is nondegenerate in the sense of Definition 4 and let us return to the investigation of the matrix elements (8.10). In view of Theorem 6, we can make a choice of the numbers μ_j such that expression (8.12) will be zero for all ε. This means that the matrix elements (8.10) will not change with changes in ε when the vector ω varies with ε according to the law $\omega = \omega_0 + \varepsilon\omega_1$.

It is not difficult to see that in this case the operator Q and, together with it, the operator function $\Psi(t)$ defined by formula (8.4) will depend on ε, but for small ε they will be uniformly bounded. This property will be important in what follows.

It is important to note that in the nondegenerate case Theorem 1 remains valid with the obvious replacement of the operator B_0 by $B_0 + D$, where D is a diagonal operator in the basis $\{a_j\}$. Inequality (4.3) now takes the form

$$|\rho(\varepsilon, \omega_0 + \varepsilon\omega_1) - \varepsilon\rho_1(\omega_1)| < C\varepsilon^{1+\delta}, \tag{9.1}$$

where ω_1 is an arbitrary frequency vector, $\rho_1(\omega_1)$ is the largest real part of the eigenvalues of the operator $B_0 + D$,* and C and δ are positive constants that can be chosen to be uniform with respect to changes in the vector ω_1 within any bounded set.

The introduction of the operator D in the averaged equation allows us to take account of the dependence of the function $\rho(\varepsilon, \omega)$ on the vector ω_1 in an explicit manner. In turn, this makes it possible for us to obtain effective formulas for finding, in parameter space $\{\varepsilon, \omega\}$, the set of given exponential growth of the solutions of Eq. (6.1) when the quantity ε is sufficiently small. This approach leads to a generalization of the corresponding formulas obtained by other methods in [9, 10, 23, 24] for equations with periodic operator coefficients.

THEOREM 7. *Let us assume that the conditions of Theorem 3 hold for the Hamiltonian Equation (6.1) for a nondegenerate frequency vector $\omega = \omega_0$. Let μ_j, with $\mu_j = \mu_j(\omega_1)$, denote the numbers that satisfy system (8.19) for a given vector ω_1 and let Q denote the set of vectors ω_1 for which the equation in λ*

$$\det\left\|\left\{\sum_k (H^{(k)} P_2 a_l, P_2 a_m)\,\mathrm{sign}\, m + [\mu_l(\omega_1) + i\lambda]\delta_{lm}\right\}\delta[(\omega_0, k) + \lambda_l - \lambda_m]\right\| = 0 \tag{9.2}$$

has roots with positive real parts.

Then, the domain of essential instability adjoining the point $\{0, \omega_0\}$ is the union of all rays of the form

$$\{\varepsilon, \omega_0 + \varepsilon\omega_1\}, \qquad \omega_1 \in \Omega.$$

Let $\varepsilon(\alpha, \omega_1)$ denote the perturbation amplitude required for the attainment of the set of α-exponential growth of the solutions of Eq. (6.1) along the ray $\{\varepsilon, \omega_0 + \varepsilon\omega_1\}$, with $\omega_1 \in \Omega$ emerging from the point $\{0, \omega_0\}$. Then, we have

$$\varepsilon(\omega_1) = \frac{\alpha}{\rho_1(\omega_1)} + O(\alpha^{1+\delta}), \tag{9.3}$$

where $\rho_1(\omega_1)$ is the largest real part of the roots of Eq. (9.2) and δ a positive number related to the order N of Eq. (9.2) by the inequality $\delta > 1/N$.

---

*It is assumed that the eigenvalues μ_j of the operator D are solutions of system (8.19).

In formula (9.2)

$$\delta_{ij} = \begin{cases} 1, & i=j, \\ 0, & i \neq j, \end{cases} \quad \delta(s) = \begin{cases} 1, & s=0, \\ 0, & s \neq 0, \end{cases}$$

and P_2 is the orthoprojector that appears in Theorem 3.

The proof is very simple. Indeed, the numbers λ defined by Eq. (9.2) are the eigenvalues of the operator $B_0 + D$ since the determinant involves the matrix generated in the basis $\{a_j\}$ by the operator $-i(B_0 + D - \lambda I)$. But, in view of the choice of the numbers $\mu_j(\omega_1)$, the largest real part of the eigenvalues of the operator $B_0 + D$ coincides with $\rho_1(\omega_0 + \varepsilon\omega_1) = \varepsilon\rho_1(\omega_1)$. Recalling Definition 3 and Theorem 1, we obtain the proof of the first part of the theorem. The proof of the second part follows from the fact that the perturbation amplitude $\varepsilon(\omega_1)$ needed for the attainment of the set of α-exponential growth along the ray $\{\varepsilon, \omega_0 + \varepsilon\omega_1\}$ satifies the relation

$$\rho(\varepsilon, \omega_0 + \varepsilon\omega_1) = \varepsilon\rho_1(\omega_1) + O(\varepsilon^{1+\delta}) = \alpha,$$

the solution of which for α yields relation (9.3). The theorem has been proved.

Remark. An analog of Theorem 7 can also be formulated for quasi-Hamiltonian equations when the matrix corresponding to the operator B_0 has spectral points in the right half-plane, the latter being defined by a finite-dimensional block of the matrix. A situation of this type occurs in the case of systems with "friction."

The order of the determinant (9.2) is in practice frequently less than or equal to two. In this case, the formulas for the set of essential instability of the solutions of Eq. (6.1) can be written down in explicit form. In view of the importance of this concept we introduce the following definition.

Definition 5. We will say that the "general" case holds for Eq. (6.1) with $\varepsilon = 0$ and $\omega = \omega_0$ if the following conditions are satisfied:

1) There exist eigenvalues λ_l, λ_m of the operator $iJ^{-1}H_0$ satisfying for an integer vector $k \in K$ the relations

$$(\omega_0, k) = \lambda_m - \lambda_l, \qquad \lambda_m > 0, \quad \lambda_l < 0. \tag{9.4}$$

2) The vector $k = k_{lm}$ and the numbers λ_m, λ_l are uniquely defined by relations (9.4).*

3) The following inequalities hold for all eigenvalues λ_p that are commensurable with λ_m and λ_l modulo (ω_0, K)†:

$$|\lambda_p - \lambda_m| > \delta, \quad |\lambda_l - \lambda_p| > \delta, \tag{9.5}$$

where $\delta = \max\limits_{k \in K} |(\omega_0, k)|$.

It should be noted that as the number δ is finite, in order to determine whether the conditions for the "general" case hold we need only examine a finite number of eigenvalues of the operator $iJ^{-1}H_0$.

---

*That is, firstly, for given $\lambda_m > 0$ and $\lambda_l < 0$ there exists a unique vector k, with $k \in K$, that satisfies the first relation of (9.4) and, secondly, the pair of numbers λ_m, λ_l that can satisfy relations (9.4) for some $k \in K$ is unique.

†A number λ_p is commensurable with λ_l modulo (ω_0, K) if we can find a vector $k \in K$ such that $\lambda_p = \lambda_l$ modulo (ω_0, K). It should be recalled that the set K of integer vectors corresponding to nonzero coefficients $H^{(k)}$ in the sum (1.2) is by definition finite.

THEOREM 8. Let us assume that the general case obtains for the Hamiltonian equation (6.1) and that either the conditions of Theorem 3 are satisfied or the Fourier indices of the operator function $B(t) = e^{J^{-1}H_0 t} J^{-1} H_0(t, \omega_0) e^{-J^{-1}H_0 t}$ do not have zero as a point of condensation. Then, the following assertions hold:

1. The set of essential instability of solutions adjoining the point $\{0, \omega_0\}$ is a cone

$$\Omega = \{\varepsilon, \omega_0 + \varepsilon\omega_1\}, \tag{9.6}$$

where ω_1 is any vector satisfying the inequalities

$$\gamma_{ll} + \gamma_{mm} - 2|\gamma_{lm}| < (\omega_1, k_{lm}) < \gamma_{ll} + \gamma_{mm} + 2|\gamma_{lm}|. \tag{9.7}$$

Here, we have

$$\gamma_{lm} = (H^{(k_{lm})}a_l, a_m), \quad H^{(k_{lm})} = \lim_{T\to\infty} \frac{1}{2T} \int_{-T}^{T} H(t, \omega_0) e^{-i(k_{lm}, \omega_0)t} dt, \tag{9.8}$$

and the integer vector k_{lm} is defined by condition (9.4).

2. Let $\varepsilon = \varepsilon(\alpha, \omega_1)$ denote the perturbation amplitude necessary for the attainment of the set of α-exponential growth along the ray $\{\varepsilon, \omega_0 + \varepsilon\omega_1\}$. If $|\gamma_{lm}| \neq 0$, then we have

$$\varepsilon(\alpha, \omega_1) = \frac{2\alpha}{\sqrt{4|\gamma_{lm}|^2 - [\gamma_{ll} + \gamma_{mm} - (\omega_1, k_{lm})]^2}} + O(\alpha^2). \tag{9.9}$$

3. If $\varepsilon(\alpha)$ is the exact upper bound to the values of ε such that the estimate

$$\|x(t)\| = O(e^{\alpha t}) \tag{9.10}$$

holds for the general solution $x(t)$ of Eq. (6.1) in the domain of instability, then we have

$$\varepsilon(\alpha) = \frac{\alpha}{|\gamma_{lm}|} + O(\alpha^2). \tag{9.11}$$

4. If the "general" case does not hold for the Hamiltonian equation (6.1) with $\varepsilon = 0$ and $\omega = \omega_0$, but there exist s pairs of numbers $\lambda_{m_p}, \lambda_{l_p}$, with $p = 1, 2, \ldots, s$, of opposite sign satisfying relations (9.4) and (9.5) and there are no identical numbers among the various pairs, then in the first approximation the domain of dynamic instability adjoining the point $\{0, \omega_0\}$ is the union in the set-theoretic sense of the s cones (9.6) constructed as indicated in 1 above, for each pair of numbers $\lambda_{m_p}, \lambda_{l_p}$.

The quantities $\varepsilon(\alpha, \omega_1)$ and $\varepsilon(\alpha)$ are in this case obtained by minimization over all s cones being considered.

To prove the theorem, we will only note that in the "general" case we have

$$\mu_m(\omega_1) = \mu_l(\omega_1) + (\omega_1, k_{lm})$$

and Eq. (9.2) for λ reduces to a second-order equation. The condition that this equation has roots in the right half-plane yields relations (9.6)-(9.8). The quantity $\varepsilon(\alpha, \omega_1)$ is defined to be the solution of the equation

$$\rho(\varepsilon, \omega_0 + \varepsilon\omega_1) = \alpha. \tag{9.12}$$

In the "general" case, we have

$$\rho(\varepsilon, \omega_0 + \varepsilon\omega_1) = \varepsilon\rho_1(\omega_1) + O(\varepsilon^2), \tag{9.13}$$

where $\rho_1(\omega_1)$ is the largest real part of the roots of Eq. (9.2) and the term $O(\varepsilon^2)$ arises because of the primality of the elementary divisors of this equation (see Theorem 1). From relations (9.2) and (9.3) we obtain

$$\varepsilon(\alpha, \omega_1) = \frac{\alpha}{\rho_1(\omega_1)} + O(\alpha^2)$$

and a simple derivation involving the calculation of $\rho_1(\omega_1)$ yields formula (9.9). The minimization of the right-hand side of (9.9) with respect to ω_1 yields formula (9.11).

Finally, when the conditions of Part 4 of Theorem 8 are satisfied, the matrix whose determinant appears in expression (9.2) can be represented as a quasi-diagonal matrix with blocks of order two, so that the left hand side of Eq. (9.2) reduces to the product of quadratic polynomials in λ.

We then repeat the arguments given during the proof of the preceding parts of the theorem.

By constructing the function $\varepsilon(\alpha, \omega_1)$ (see Part 2 of Theorem 8) for various values of ω_1 for which $|\gamma_{lm}| \neq 0$, we can determine the boundary of the domain of α-exponential growth to within a small quantity of the second order in ε.

It follows from Theorem 8 that with increasing values of

$$|\gamma_{lm}| = |(H^{(k_{lm})}a_l, a_m)|, \tag{9.14}$$

on the one hand, the "opening" of the corresponding domain of dynamic instability in the first approximation becomes wider and, on the other hand, the set of α-exponential growth approaches for small α the point $\{0, \omega_0\}$. This property is a characteristic feature of Hamiltonian systems; this situation does not occur in general in the case of non-Hamiltonian systems. In the case of Hamiltonian systems, the quantity (9.14) can thus be taken as a measure of the "danger" due to the resonance frequency vector ω_0. The analogous fact for systems with periodic coefficients is well known [6-10].

Let us make several remarks that amplify Theorem 8. It follows from the definition of the "general" case that formulas (9.7)-(9.9) were derived without any special assumptions concerning the properties of the matrix elements (9.8). It is sometimes found in practice that some of the matrix elements $(H^{(k)}a_l, a_m)$ become zero because of some of the properties of the Fourier coefficients $H^{(k)}$ of the function $H(t, \omega)$. Because of this, Eq. (9.2) in λ reduces in a number of cases to a second-order equation even without any need to assume the presence of the "general" case. This situation occurs, for example, in the case of some reducible systems. It is clear that the formulas of Theorem 7 are also valid in this case. In particular, this leads to the following result.

Let us assume that the class of perturbations $\varepsilon H(t, \omega)$ in Eq. (6.1) coincides with the class of all symmetric functions $H(t, \omega)$ of the form of (1.2). We can then assert that if the vector ω_0 satisfies relation

(9.4) with some integer vector k and some eigenvalues λ_l, λ_m of the operator $iJ^{-1}H_0$ of opposite signs, then ω_0 is a resonance vector. Indeed, we can choose a symmetric operator function $H(t, \omega_0)$ of the form of (1.2) such that the corresponding Eq. (9.2) will reduce to a second-order equation, the investigation of which leads to formulas (9.9). Thus, in order for the vector ω_0 to be a resonance vector in the class of all Hamiltonian equations of the type being considered, it is necessary and sufficient that it satisfy a relation of the form of (9.4) with numbers λ_m and λ_l of opposite signs. This argument serves as the proof of the last part of Theorem 5.

Analogs of Theorems 7 and 8 can also be formulated for quasi-Hamiltonian equations. We will restrict ourselves to the case when the operator $B_0 + D$ of the averaged Eq. (8.8) is essentially two-dimensional.

THEOREM 9. Let us assume that the following conditions hold for the quasi-Hamiltonian equation (6.1) and a frequency vector $\omega = \omega_0$:

1) For an arbitrary sufficiently small positive δ, the inequality

$$|(\omega_0, k) - \lambda_p + \lambda_{p'}| < \delta,$$

where k runs over the elements of the set K, $k \neq 0$, and λ_p, $\lambda_{p'}$ are all possible eigenvalues of the operator $iJ^{-1}H_0$, is satisfied by a unique pair of numbers λ_m, λ_l, the vector $k = k_{lm}$ being uniquely defined by the condition $(\omega_0, k_{lm}) = \lambda_m - \lambda_l$.

2) The matrix elements $(H^{(k)} a_l, a_m)$ are real numbers.*

3) The operator $iPH^{(0)}P$, where P is the orthoprojector on an arbitrary characteristic subspace of the operator $iJ^{-1}H_0$, does not possess eigenvalues in the right half-plane.†

If in the notation of Theorem 8 the following inequality holds:

$$\gamma_{lm}\gamma_{ml} \operatorname{sign} m \cdot \operatorname{sign} l \equiv -\Delta_{lm} < 0, \tag{9.15}$$

then all of the assertions of the theorem are valid with the following modifications:

1. Inequalities (9.7) are replaced by

$$\gamma_{ll} \operatorname{sign} l - \gamma_{mm} \operatorname{sign} m - \sqrt{\Delta_{lm}} < (\omega_1, k_{lm}) < \gamma_{ll} \operatorname{sign} l - \gamma_{mm} \operatorname{sign} m + \sqrt{\Delta_{lm}}. \tag{9.16}$$

2-3. Formulas (9.9) and (9.11) are replaced by

$$\varepsilon(\alpha, \omega_1) = \frac{2\alpha}{\sqrt{4\Delta_{lm} - [\gamma_{ll} \operatorname{sign} l - \gamma_{mm} \operatorname{sign} m - (\omega_1, k_{lm})]^2}} + O(\alpha^2), \tag{9.17}$$

$$\varepsilon(\alpha) = \frac{\alpha}{\sqrt{\Delta_{lm}}} + O(\alpha^2). \tag{9.18}$$

4. The requirement that $\lambda_{m_p} \cdot \lambda_{m_l} < 0$ may be omitted.

---

*This condition is usually satisfied in practice.

†This condition is obviously satisfied either if $H^{(0)}$ is the null operator or if the eigenvalues λ_j of the operator $iJ^{-1}H_0$ are simple (see the preceding condition).

The proof of this theorem is simply a matter of several obvious transformations. In the "general" case for a Hamiltonian equation, conditions 1)-3) of Theorem 9 are satisfied, moreover $\gamma_{lm}\gamma_{ml} = |\gamma_{lm}|^2$, and condition (9.15) can only be satisfied of sign m = −sign l. Relations (9.16)-(9.18) now become the relations of Theorem 8. In the case of non-Hamiltonian equations, relation (9.15) can also be satisfied when sign l = sign m (for example, see Section 4 of [10]).

The same remarks can be made in the case of Theorem 9 as in the case of Theorem 8. In particular, we can assert that the vector ω_0 is a resonance vector in the class of all quasi-Hamiltonian equations if and only if it satisfies a relation of the form of (9.4) with an integer vector k and eigenvalues λ_l, λ_m of the operator $iJ^{-1}H_0$. It is obvious that the class of perturbations for which this assertion is correct may be narrowed down to a large extent, but this question will not be discussed here.

§ 10. The Case of Degeneracy

The investigation of degeneracy presents considerable difficulties and can be carried out fully only in special cases. Some of these have already been discussed above in principle. In fact, Theorem 5 was formulated before the introduction of the concept of degeneracy and, consequently, covers some of these cases.

On the assumption that there is degeneracy in the vector $\omega = \omega_0$, let us perform the transformation of Eq. (4.1) in exactly the same manner as was done in the proof of Theorem 3. Let us restrict ourselves for simplicity to the case in which the Fourier indices of the function

$$B(t) = e^{-J^{-1}H_0t}J^{-1}H(t, \omega_0)\,e^{J^{-1}H_0t} \tag{10.1}$$

do not possess a condensation point at zero.*

Let us now assume that the perturbation-frequency vector is given by

$$\omega = \omega_0 + \varepsilon\omega_1, \tag{10.2}$$

where ω_1 is an arbitrary fixed vector. It is obvious that the Fourier indices of the function B(t) will now vary continuously as ε varies. Let us select and fix a positive number δ such that for any ω_1 belonging to a bounded set and a sufficiently small ε the function $B(t) = B(t, \varepsilon)$ has in the δ-neighborhood of a zero only those Fourier indices that become zero when $\omega_1 = 0$. B(t) can then be represented as

$$B(t) = D_1(t, \varepsilon) + D_2(t, \varepsilon), \tag{10.3}$$

where for small ε the function $D_1(t, \varepsilon)$ has Fourier indices only outside the δ-neighborhood of zero and $D_2(t, \varepsilon)$ has Fourier indices in the δ-neighborhood of zero. For these values of ε, the function

$$\Psi_1(t) = \int_0^t D_1(s, \varepsilon)\,ds \tag{10.4}$$

will obviously be a bounded operator from **H** into **C**, while the change of variable

$$y(t) = (I + \varepsilon\Psi_1(t))^{-1}\,e^{-J^{-1}H_0t}x(t) \tag{10.5}$$

* If the conditions of Theorem 3 are satisfied, then instead of the function B(t) we should consider the function $P_2B(t)P_2$ which has only a finite number of Fourier indices.

will transform Eq. (6.1) into the equation

$$\frac{dy}{dt} = \varepsilon D_2(t, \varepsilon) y + \varepsilon^2 D_3(t, \varepsilon) y, \tag{10.6}$$

where $D_3(t, \varepsilon)$ for all sufficiently small ε is an operator function uniformly bounded with respect to t. The Fourier indices of $D_2(t, \varepsilon)$ are of the form $i\varepsilon(\omega_1, k)$, so that the equation

$$\frac{dy}{dt} = \varepsilon D_2(t, \varepsilon) y \tag{10.7}$$

is an equation with slowly varying coefficients. Introducing the slow time $\tau = \varepsilon t$ and writing $\tilde{D}_2(\tau) = D_2\left(\frac{\tau}{\varepsilon}, \varepsilon\right)$, $\tilde{D}_3(\tau, \varepsilon) = D_3\left(\frac{\tau}{\varepsilon}, \varepsilon\right)$, we obtain instead of (9.4) the equation

$$\frac{dy}{d\tau} = [\tilde{D}_2(\tau) + \varepsilon \tilde{D}_3(\tau, \varepsilon)] y. \tag{10.8}$$

The operator function $D_2(\tau)$ no longer depends on ε, while $\tilde{D}_3(\tau, \varepsilon)$ is uniformly bounded with respect to τ for small values of ε.

The index $\rho_1(\varepsilon, \omega)$ of the exponential growth of solutions of Eq. (10.8) is related to the index $\rho(\varepsilon, \omega)$ of the exponential growth of solutions of Eq. (6.1) by the obvious equation

$$\varepsilon \rho_1(\varepsilon, \omega) = \rho(\varepsilon, \omega)$$

and, therefore, to find the latter it is sufficient to know how to calculate $\rho_1(\varepsilon, \omega)$. In the given case, however, the truncated equation

$$\frac{dy}{d\tau} = \tilde{D}_2(\tau) y$$

depends on τ and, in general, cannot be integrated analytically. In those case when Eq. (10.9) can be integrated analytically, the derivation of effective formulas for $\rho_1(\varepsilon, \omega)$ for small ε does not present any special difficulties in general.

In the general case, system (10.9) is almost periodic with a basis whose form and dimensions depend on the choice of the vector ω_1. Using the transformation given in Section 8, we can attempt to decrease the number of the Fourier indices of the coefficients of the truncated equation for a given vector ω_1. The following assertion holds.

THEOREM 10. Let the frequency vector ω_0 be degenerate and let the rank of its degeneracy be equal to r. By a suitable choice of the eigenvalues $i\mu_j$ of the operator D in expression (8.7) we can obtain a transformation (8.3) which leads to an averaged equation of the form of (10.8) in which the number of nonzero Fourier indices of the operator function $\tilde{D}_2(\tau)$ does not exceed r.

Proof. In view of Theorem 5, for any vector ω_1 we can choose the μ_j such that all relations of system (8.19) are satisfied with the possible exception of some r relations. It is obvious that only these unsatisfied relations can lead to the appearance of nonzero Fourier indices of the function $\tilde{D}_2(\tau)$, as can be seen from the transformation (10.5) being carried out. This proves the theorem.

In practice, the set K of integer vectors indexing the Fourier coefficients of the function $H(t, \omega_0)$ is symmetric with respect to the coordinate origin, i.e., in addition to $k \in K$ we also have $-k \in K$. Because of this, in addition to the vector $\sum_{s=1}^{l} m_s^{(p)} k_s$, if it is different from the zero vector, the system (8.16) must also involve the vector $-\sum_{s=1}^{l} m_s^{(p)} k_s$. Consequently, the rank of degeneracy can only be an even number and the Fourier indices of the operator function $\tilde{D}_2(\tau)$ are symmetric with respect to zero. In cases that are of greatest practical interest, the rank of degeneracy of the frequency vector ω_0 is equal to two.*

Theorem 10 in this case leads to an important corollary.

Corollary. If the rank of degeneracy of the frequency vector ω_0 is equal to two, then transformation (10.5) leads to an averaged equation (10.8) with a periodic operator function $\tilde{D}_2(\tau)$.

Although equations with periodic coefficients as a rule are not integrable in closed form, the investigation of the stability of the solutions of such equations is a comparatively easier task than the corresponding investigation in the case of equations whose coefficients are quasi-periodic. The Floquet–Lyapunov theorem holds for equations with periodic coefficients and, in particular, it follows from this theorem that the averaged equation can be put into a form in which the principal part is independent of the time. Therefore, for investigating degeneracy of the second rank we can use the assertions that were derived for averaged equations with a time independent principal part. It is clear that the derivation of effective formulas for the exponential-growth index depends on the possibility of the solution of the truncated equation in explicit form.

Literature Cited

1. V. V. Bolotin, Dynamic Stability of Elastic Systems [in Russian], GITTL, Moscow (1956).
2. E. A. Beilin and G. Yu. Dzhanelidze, "A survey of articles on the dynamic stability of elastic systems," Priklad. Matem. i Mekhan., Vol. 16, No. 5 (1952).
3. I. I. Gol'denblat, Dynamic Stability of Structures [in Russian], GITTL, Moscow (1956).
4. M. G. Krein and V. A. Yakubovich, "Hamiltonian systems of linear differential equations with periodic coefficients," in: Transactions of the International Symposium on Nonlinear Oscillations, Vol. 1 [in Russian], Izd. KGU, Kiev (1963).
7. V. A. Yakubovich and V. M. Starzhinskii, "Parametric resonance in systems with many degrees of freedom," in: Transactions of the Interuniversity Conference on Applied Stability of Motion and Analytic Mechanics [in Russian], Izd. KazGU, Kazan' (1964).
8. K. G. Valeev, "On the dangerousness of combinative resonances," Priklad. Matem. i Mekhan., Vol. 27, No. 6 (1963).
9. V. N. Fomin, "Parametric resonance in elastic systems with an infinite number of degrees of freedom," Part I, Vestnik LGU, No. 13 (1965); Part II, Vestnik LGU, No. 19 (1965).
10. V. N. Fomin, "Dynamic instability domains of parametrically excited systems with an infinite number of degrees of freedom," in: V. I. Smirnov (editor), Problems in Mathematical Analysis, Vol. 1, Consultants Bureau, New York (1968).
11. E. Hille and R. Phillips, Functional Analysis and Semi-Groups, American Mathematical Society, Providence (1957).
12. V. I. Derguzov, "Continuous dependence of the maximum index of exponential growth of the solution of linear Hamiltonian equations with periodic operator coefficients," in: V. I. Smirnov (editor), Problems in Mathematical Analysis, Vol. 1, Consultants Bureau, New York (1968).

*If the set K does not contain a vector orthogonal to the frequency vector ω_0, then degeneracy of rank two can be considered to be the general case with respect to all cases of degeneracy. Indeed, the higher the rank of degeneracy of the frequency vector ω_0, the greater the number of relations that must be satisfied by its components.

13. M. G. Krein, Lectures on the Theory of Stability of Solutions of Differential Equations in Banach Spaces [in Russian], Izd. AN Ukrainsk. SSR, Kiev (1964).
14. V. I. Derguzov, "On the stability of solutions of Hamilton's equations with unbounded periodic operator coefficients," Matem. Sbornik, Vol. 63 (105), 4 (1964).
15. I. Z. Shtokalo, Linear Differential Equations with Variable Coefficients [in Russian], Izd. AN Ukrainsk. SSR, Kiev (1960).
16. K. G. Valeev, "An investigation of the stability of a quasistationary system of linear differential equations with almost-periodic coefficients," Izv. Vuzov. Radiofizika, Vol. 5 (1962).
17. N. M. Krylov and N. N. Bogolyubov, Introduction to Nonlinear Mechanics [in Russian], Izd. AN Ukrainsk. SSR, Kiev (1937).
18. N. N. Bogolyubov and Yu. A. Mitropol'skii, Asymptotic Methods in the Theory of Nonlinear Oscillations, Fizmatgiz, Moscow (1958).
19. B. M. Levitan, Almost-Periodic Functions [in Russian], Gostekhizdat, Moscow (1953).
20. F. Reisz and B. S. Nagy, Lectures on Functional Analysis [Russian translation], IL, Moscow (1956).
21. K. P. Persidskii, "On the characteristic numbers of differential equations," Izv. AN KazSSR, Seriya Matem. i Mekhan., No. 1 (1947).
22. M. G. Krein, "The fundamentals of the theory of λ-zones of stability of a canonical system of linear differential equations with periodic coefficients," in: In Memory of A. A. Andronov [in Russian], Izd. AN SSSR, Moscow (1955).
23. V. N. Fomin and V. A. Yakubovich, "The calculation of the characteristic indices of linear systems with periodic coefficients," in: Computational Methods, Vol. 3 [in Russian], Izd. LGU (1966).
24. V. N. Fomin, "Parametric resonance in elastic systems with distributed parameters," Dokl. Akad. Nauk SSSR, Vol. 164, No. 1 (1965).

SOME PROPERTIES OF TRANSFORMERS DEFINED BY DOUBLE-INTEGRAL OPERATORS

K. Töllner

§ 1. Introduction

The present article is devoted to the examination of several problems associated with double-integral operators of the form

$$\int_{\Lambda}\int_{M} \varphi(\lambda, \mu) F(d\mu) TE(d\lambda). \tag{1.1}$$

Here, $E(\cdot)$ and $F(\cdot)$ are two orthogonal spectral measures defined in separable Hilbert space H on σ-algebras in sets Λ and M, respectively, T is a bounded linear operator in H, and $\varphi(\lambda, \mu)$ is a complex valued function defined on the set $\Lambda \times$ M. The definition of the integral (1.1) has been given in [1, 2]. These articles also contain a discussion of its principal properties and an account of various applications.

If integral (1.1) is considered as a function of the operator T, then we obtain a linear operator (transformer) Φ acting on operators in H. We will study the properties of Φ in symmetrically normed ideals (s.n. ideals) of the ring **R** of all bounded linear operators in H. Let $\mathbf{S}_{\rm I}$ and $\mathbf{S}_{\rm II}$ be two s.n. ideals of the ring **R**. For some functions $\varphi(\lambda, \mu)$, the integral (1.1) defines a bounded linear transformer Φ acting from the s.n. ideal $\mathbf{S}_{\rm I}$ into the s.n. ideal $\mathbf{S}_{\rm II}$. The collection of all such transformers is a normed linear space $(\mathbf{S}_{\rm I}, \mathbf{S}_{\rm II})$ (or a commutative normed ring $r(\mathbf{S}) = (\mathbf{S}, \mathbf{S})$ when $\mathbf{S}_{\rm I} = \mathbf{S}_{\rm II} = \mathbf{S}$). Spaces $(\mathbf{S}_{\rm I}, \mathbf{S}_{\rm II})$ for the case when $\mathbf{S}_{\rm I} \subset \mathbf{S}_2$ or $\mathbf{S}_{\rm I} \supset \mathbf{S}_2$ (here, $\mathbf{S}_2$ is the ideal of all Hilbert–Schmidt operators) have been considered in [2] (see also [1]). The present article is devoted to spaces $(\mathbf{S}_{\rm I}, \mathbf{S}_{\rm II})$ for a wider class of s.n. ideals: $\mathbf{S}_{\rm I}$ will include arbitrary separable s.n. ideals. The general definition of the space $(\mathbf{S}_{\rm I}, \mathbf{S}_{\rm II})$ is given in Section 2.

The first result of the present article is obtained in Section 3. The following lower bound to the norm of transformer Φ in space $(\mathbf{S}_{\rm I}, \mathbf{S}_{\rm II})$ in terms of the uniform norm of the function $\varphi(\lambda, \mu)$ is obtained there:

$$\|\Phi\|_{\mathbf{S}_{\rm I}, \mathbf{S}_{\rm II}} \geqslant \operatorname*{vrai\,sup}_{\lambda \in \Lambda, \mu \in M} |\varphi(\lambda, \mu)|. \tag{1.2}$$

(The supremum is taken with respect to the measure indicated in Section 2.) This result has been obtained in [2] for the special case $\mathbf{S}_{\rm I} = \mathbf{S}_{\rm II} \subset \mathbf{S}_2$. The completeness of the space $(\mathbf{S}_{\rm I}, \mathbf{S}_{\rm II})$ follows from (1.2).

The spaces of transformers in arbitrary s.n. ideals are investigated in Section 4. Let S be a fixed (although arbitrary) s.n. ideal. Each function of the form

$$\varphi(\lambda, \mu) = \sum_{i=1}^{n} x_i(\lambda) y_i(\mu),$$

where $x_i(\lambda)$ and $y_i(\mu)$ are continuous functions on Λ and M, respectively, is associated with a continuous linear transformer Φ_S given by

$$\Phi_S S = \sum_{i=1}^{n} Y_i S X_i \quad (S \in \mathbf{S}),$$

where

$$X_i = \int_{\Lambda} x_i(\lambda) E(d\lambda), \quad Y_i = \int_{M} y_i(\mu) F(d\mu) \quad (i = 1, \ldots, n).$$

The totality of these transformers for all functions of the above type forms a normed ring. We will consider its completion m(**S**). Moreover, if the s.n. ideal **S** is contained in $\mathbf{S}_2$ or is separable, then the transformers belonging to m(**S**) are defined by the integral (1.1) and m(**S**) is found to be a subring of the ring r(**S**) = (**S**, **S**). Some of the properties of m(**S**) are studied in Section 4. Firstly, the maximal ideals of these rings are described. Secondly, it is proved that they are without the radical when one of the following three conditions are satisfied: 1) $\mathbf{S} \subset \mathbf{S}_2$, 2) the ideal **S** is separable, and 3) the ideal **S** is conjugate to a separable s.n. ideal. In addition, the presence of natural involution in rings m(**S**) is established in this case.

The concepts of the theory of s.n. ideals used in this article can be found in Chapter 3 of [3]. In addition, some basic concepts of the general theory of commutative normed rings will also be needed (see Chapter 1 of [4]).

§ 2. The Basic Constructions

This section contains a brief description of the construction of integral (1.1) which was given in [1, 2]. The definition of the space ($\mathbf{S}_{\mathrm{I}}$, $\mathbf{S}_{\mathrm{II}}$) follows.

Let us begin by introducing the notation that will be used everywhere in the following. Let H, $E(\cdot)$, $F(\cdot)$, Λ, and M have the meaning given them in Section 1. This meaning will remain fixed throughout the article. The letter **R** denotes the ring of all continuous linear operators in H; $\|S\|$ is the norm of an operator S belonging to **R**. Symmetrically normed ideals of ring **R** are denoted by **S**; $\|S\|_{\mathbf{S}}$ is the norm of an operator S belonging to **S**. The ideal of all Hilbert–Schmidt operators is denoted by $\mathbf{S}_2$. As is well known, $\mathbf{S}_2$ is a Hilbert space with the scalar product $\langle S, T\rangle = \mathrm{sp}\,(ST^*)$ $(S, T \in S_2)$.

Let A and B be two Banach spaces; the space of all continuous linear operators mapping A into B is denoted by [A, B]. The norm in space [A, B] is denoted by $\|\cdot\|_{A, B}$.

Let us now proceed to the construction of the integral (1.1). The symbol $\delta(\partial)$ denotes the subsets of the set Λ (M) which are measurable in the spectral measure $E(\cdot)(F(\cdot))$, and Δ denotes the rectangle $\Delta = \delta \times \partial$.

For each rectangle $\Delta = \delta \times \partial$, we define a continuous linear transformer G(Δ) in Hilbert space $\mathbf{S}_2$,

$$G(\Delta) S = F(\partial) S E(\delta) \quad (S \in \mathbf{S}_2). \tag{2.1}$$

The rectangle function $G(\Delta)$ can be extended in the standard manner to an orthogonal spectral measure $G(\cdot)$ defined on a σ-algebra of subsets of $\Lambda \times M$.

Let a complex-valued function $\varphi(\lambda, \mu)$ be defined on $\Lambda \times M$, the function being essentially bounded in measure $G(\cdot)$. We can then assign a meaning to the integral operator

$$\Phi = \int_{\Lambda}\int_{M} \varphi(\lambda, \mu)\, dG(e), \tag{2.2}$$

which defines a continuous linear transformer Φ in $\mathbf{S}_2$. In addition, we have

$$\|\Phi\|_{\mathbf{S}_2, \mathbf{S}_2} = G - \sup|\varphi(\lambda, \mu)|. \tag{2.3}$$

Let us give a definition of the integral (1.1). For all operators S belonging to $\mathbf{S}_2$ we take

$$\Phi S \equiv \int_{\Lambda}\int_{M} \varphi(\lambda, \mu) F(d\mu)\, SE(d\lambda).$$

The totality of transformers Φ of the form of (2.2) is a subring $(\mathbf{S}_2, \mathbf{S}_2)$ of the ring $[\mathbf{S}_2, \mathbf{S}_2]$. Associating transformers Φ with the functions $\varphi(\lambda, \mu)$, we obtain an isometric isomorphism of the ring of all essentially bounded (in measure G) functions onto the ring $(\mathbf{S}_2, \mathbf{S}_2)$.

We will also consider the transformer Φ in s.n. ideals different from $\mathbf{S}_2$. Let $\mathbf{S}_{\mathrm{I}}$ and $\mathbf{S}_{\mathrm{II}}$ be two s.n. ideals, $\mathbf{S}_{\mathrm{I}}$ being either contained in $\mathbf{S}_2$ or separable. In both cases, the intersection $\mathbf{S}_{\mathrm{I}} \cap \mathbf{S}_2$ is dense in $\mathbf{S}_{\mathrm{I}}$. Let the transformer $\Phi \in (\mathbf{S}_2, \mathbf{S}_2)$ map the intersection $\mathbf{S}_{\mathrm{I}} \cap \mathbf{S}_2$ into $\mathbf{S}_{\mathrm{II}}$ and let this mapping be continuous with respect to the norms $\|\cdot\|_{\mathbf{S}_{\mathrm{I}}}$ and $\|\cdot\|_{\mathbf{S}_{\mathrm{II}}}$, i.e.,

$$\|\Phi S\|_{\mathbf{S}_{\mathrm{II}}} \leqslant c\|S\|_{\mathbf{S}_{\mathrm{I}}} \qquad (S \in \mathbf{S}_{\mathrm{I}} \cap \mathbf{S}_2).$$

$\Phi|_{\mathbf{S}_{\mathrm{I}} \cap \mathbf{S}_2}$ can then be extended in continuity to the transformer $\Phi_{\mathbf{S}_{\mathrm{I}}, \mathbf{S}_{\mathrm{II}}} \in [\mathbf{S}_{\mathrm{I}}, \mathbf{S}_{\mathrm{II}}]$. In this case, we say that the function $\varphi(\lambda, \mu)$ generates the transformer $\Phi_{\mathbf{S}_{\mathrm{I}}, \mathbf{S}_{\mathrm{II}}}$.

Definition. The collection of all transformers $\Phi_{\mathbf{S}_{\mathrm{I}}, \mathbf{S}_{\mathrm{II}}}$ generated by the functions $\varphi(\lambda, \mu)$ will be called the class $(\mathbf{S}_{\mathrm{I}}, \mathbf{S}_{\mathrm{II}})$.

The mapping $\varphi \to \Phi_{\mathbf{S}_{\mathrm{I}}, \mathbf{S}_{\mathrm{II}}}$, is obviously linear. When $\mathbf{S}_{\mathrm{I}} = \mathbf{S}_{\mathrm{II}}$, a product of several functions generates the product of the corresponding transformers.

Distinct functions generate distinct transformers.* Indeed, in view of (2.3) we have $\Phi' \neq \Phi''$ when $\varphi' \neq \varphi''$. Since the intersection $\mathbf{S}_{\mathrm{I}} \cap \mathbf{S}_2$ is dense in $\mathbf{S}_2$, we have $\Phi'|_{\mathbf{S}_{\mathrm{I}} \cap \mathbf{S}_2} \neq \Phi''|_{\mathbf{S}_{\mathrm{I}} \cap \mathbf{S}_2}$, and this implies that $\Phi'_{\mathbf{S}_{\mathrm{I}}, \mathbf{S}_{\mathrm{II}}} \neq \Phi''_{\mathbf{S}_{\mathrm{I}}, \mathbf{S}_{\mathrm{II}}}$.

It follows from what has been said above that the class $(\mathbf{S}_{\mathrm{I}}, \mathbf{S}_{\mathrm{II}})$ is a linear subset of the space $[\mathbf{S}_{\mathrm{I}}, \mathbf{S}_{\mathrm{II}}]$ isomorphic to a linear set of functions $\varphi(\lambda, \mu)$. If we have $\mathbf{S}_{\mathrm{I}} = \mathbf{S}_{\mathrm{II}} = \mathbf{S}$, then the class $(\mathbf{S}, \mathbf{S})$ is a subring of the ring $[\mathbf{S}, \mathbf{S}]$ isomorphic to a (commutative) ring of functions $\varphi(\lambda, \mu)$. We will write $r(\mathbf{S})$ instead of $(\mathbf{S}, \mathbf{S})$.

It has been shown in [2] that a natural involution which corresponds to the transition from the function $\varphi(\lambda, \mu)$ to the complex conjugate function $\overline{\varphi(\lambda, \mu)}$ exists in spaces $(\mathbf{S}_{\mathrm{I}}, \mathbf{S}_{\mathrm{II}})$, where $\mathbf{S}_{\mathrm{I}} \subset \mathbf{S}_2$. This result and its proof can be immediately transferred to the general case.

LEMMA 2.1. *If the function $\varphi(\lambda, \mu)$ generates the transformer $\Phi_{\mathbf{S}_{\mathrm{I}}, \mathbf{S}_{\mathrm{II}}}$ of class $(\mathbf{S}_{\mathrm{I}}, \mathbf{S}_{\mathrm{II}})$, then the complex conjugate function $\overline{\varphi(\lambda, \mu)}$*

---

* Functions that are equivalent in measure G are not to be distinguished.

generates a transformer $\overline{\Phi}_{S_I, S_{II}}$ of the same class. In addition, we have

$$\|\Phi_{S_I, S_{II}}\|_{S_I, S_{II}} = \|\overline{\Phi}_{S_I, S_{II}}\|_{S_I, S_{II}}. \tag{2.4}$$

In conclusion, we show that the result of the application of a transformer to an operator depends only on the operator and on the function generating this operator, but is independent of the s.n. ideals S_I and S_{II} in which the given transformer is considered. As always, it is here assumed that the spectral measures $E(\cdot)$ and $F(\cdot)$ are fixed.

LEMMA 2.2. Let the transformers

$$\Phi_{S_I, S_{II}} \in (S_I, S_{II}) \text{ and } \Phi_{S_I', S_{II}'} \in (S_I', S_{II}')$$

be generated by the same function $\varphi(\lambda, \mu)$. Then, we have

$$\Phi_{S_I, S_{II}} S = \Phi_{S_I', S_{II}'} S \quad (S \in S_I \cap S_I').$$

Proof. Let S belong to $S_I \cap S_I'$. The assertion is trivial when S belongs to S_2. Let us therefore assume that we have $S \notin S_2$. Then, neither S_I nor S_I' are contained in S_2 and, consequently, S_I and S_I' are separable. Let us consider the partial sums S_n of the Hilbert–Schmidt series for the operator S. S_n converges to S in both the S_I and S_I' norms. Consequently, we have

$$\Phi_{S_I, S_{II}} S_n \to \Phi_{S_I, S_{II}} S \text{ in } S_{II}$$

and

$$\Phi_{S_I', S_{II}'} S_n \to \Phi_{S_I', S_{II}'} S \text{ in } S_{II}'.$$

Since S_n belongs to S_2, we have

$$\Phi_{S_I, S_{II}} S_n = \Phi_{S_I', S_{II}'} S_n = \Phi S_n \quad (n = 1, 2, \ldots).$$

If we take account of the fact that convergence in s.n. ideals implies ordinary operator convergence, then we may conclude from the above that

$$\Phi_{S_I, S_{II}} S = \Phi_{S_I', S_{II}'} S.$$

Lemma 2.2 allows us to write ΦS instead of $\Phi_{S_I, S_{II}} S$.

§ 3. The Lower Bound to the Transformer Φ

The aim of the present section is to establish inequality (1.2). This inequality has been established in [2] on the assumption that we have $S_I = S_{II} \subset S_2$. The inequality can be derived directly from one of M. G. Krein's theorems (see [5], pp. 71-72, Theorem 1.2, and the remark), although this theorem can only be applied if the condition $S_I = S_{II} \subset S_2$ is satisfied. In the present section, inequality (1.2) is proved by another method for the general case. The basis for the proof is provided by the following lemma.

LEMMA 3.1. Let $\Phi \in r(S_2)$ be a transformer generated by a function $\varphi(\lambda, \mu)$ which is essentially bounded (in measure G). We have

$$m \equiv \sup_{\substack{S \text{ one-dimensional} \\ \|S\|=1}} \|\Phi S\| = \|\Phi\|_{S_2, S_2} = G - \sup|\varphi(\lambda, \mu)| \equiv M.$$

Proof. Since the ordinary operator norm does not exceed the norm in s.n. ideals, while both norms are the same in the case of one-dimensional operators, the inequality $m \leq M$ is obviously true. Let us prove the converse in equality. Let S be a one-dimensional operator with the norm $\|S\|=1$. S can be written as

$$S=(\cdot, f)g \quad (f, g \in H;\ \|f\|=\|g\|=1).$$

Let us introduce the measures σ_f and τ_g defined by

$$\sigma_f(\delta)=\|E(\delta)f\|^2, \quad \tau_g(\partial)=\|F(\partial)g\|^2$$

[where the sets $\delta \subset \Lambda$ and $\partial \subset M$ are measurable in the measures $E(\cdot)$ and $F(\cdot)$ respectively] and the spaces

$$H_1=L_2(\sigma_f), \quad H_2=L_2(\tau_g).$$

It has been shown in [1], Section 3, that the operator ΦS is unitarily equivalent to the integral operator K with kernel $\varphi(\lambda, \mu)$ acting from H_1 into H_2 according to the formula

$$Ku=v; \quad v(\mu)=\int_\Lambda \varphi(\lambda, \mu)\, u(\lambda)\, \sigma_f(d\lambda) \quad (\mu \in M).$$

To complete the proof, we must establish that whatever our choice of $\eta > 0$, we can always find elements $f, g \in H$ (with $\|f\|=\|g\|=1$) such that the integral operator K corresponding to them has a norm given by $\|K\| \geqslant M_\eta \equiv M-\eta$.

First of all, let us clarify the basis of the proof of the last assertion. Let us assume that Λ and M are topological spaces and that the function $\varphi(\lambda, \mu)$ is continuous on the product $\Lambda \times M$. Let $|\varphi(\lambda, \mu)|$ reach its maximum value at the point (λ_0, μ_0). We assume that we have

$$\varphi(\lambda_0, \mu_0)=\max_{(\lambda, \mu) \in \Lambda \times M} |\varphi(\lambda, \mu)| \equiv M.$$

Because of the continuity of $\varphi(\lambda, \mu)$, we can find a neighborhood $\Delta = \delta \times \partial$ of (λ_0, μ_0) such that we have

$$\operatorname{Re}\varphi(\lambda, \mu) \geqslant M_\eta \quad ((\lambda, \mu) \in \Delta).$$

Choosing the elements f and g such that the measures σ_f and τ_g are normalized and concentrated on δ and ∂, we obtain the estimate $\|K\| \geqslant M_\eta$ for the norm of the corresponding operator K.

In the general case when $\varphi(\lambda, \mu)$ is an arbitrary measurable function essentially bounded (in measure G), we are unable to choose a rectangle $\Delta = \delta \times \partial$ such that we have $\operatorname{Re}\varphi(\lambda, \mu) \geqslant M_\eta$ everywhere inside it. However, we can always choose Δ to be such that this inequality holds everywhere in it with the exception of a set whose $\sigma_f \times \tau_g$ measure is small by comparison with the $\sigma_f \times \tau_g$ measure of the whole rectangle Δ. This allows us to obtain the required estimate.

Let us now give a formal proof.

1) We fix ε within the range $0 < \varepsilon < M$ and the natural number n $(n > 2)$. Let us consider the sets $A^k_{\varepsilon, n}$ defined by

$$A^k_{\varepsilon,n} = \Big\{(\lambda, \mu) \in \Lambda \times \mathrm{M} : |\varphi(\lambda, \mu)| > M_\varepsilon,$$
$$\frac{(2k-1)\Pi}{n} \leqslant \arg\varphi(\lambda, \mu) < \frac{(2k+1)\Pi}{n}\Big\} \quad (k = 0, 1, \ldots, n-1).$$

At least one of these n sets has a nonzero spectral G-measure; let it be, say, $A^{k_0}_{\varepsilon,n} = A$.

2) For arbitrary elements f, g belonging to H, we define the rectangle function

$$\rho_{f,g}(\Delta) = \sigma_f(\delta)\tau_g(\partial) \quad (\Delta = \delta \times \partial),$$

which can be extended in the standard manner to the complete measure $\rho_{f,g} \equiv \sigma_f \times \tau_g$.

The following inequality holds for all G-measurable sets A:

$$\rho_{f,g}(A) = \| G(A) S \|_2^2, \text{ where } S = (\cdot, f) g.$$

Indeed, for the rectangles $\Delta = \delta \times \partial$, we have

$$G(\Delta) S = F(\partial) S E(\delta) = (\cdot, E(\delta) f) F(\partial) g,$$

which yields

$$\| G(\Delta) S \|_2^2 = \| E(\delta) f \|^2 \| F(\partial) g \|^2 = \sigma_f(\delta)\tau_g(\partial) = \rho_{f,g}(\Delta).$$

Let us introduce the class of sets defined by

$$Z = \Big\{B \subset \Lambda \times \mathrm{M} : B = \bigcup_{i=1}^{\infty} \Delta_i;\ \Delta_i = \delta_i \times \partial_i,\ \Delta_i \cap \Delta_j = \varnothing \quad (i \neq j)\Big\}.$$

For any G-measurable set A and any operator S belonging to $\mathbf{S}_2$, we have (see [6], Section 14)

$$\| G(A) S \|_2^2 = \inf_{\substack{B \in Z \\ B \supset A}} \| G(B) S \|_2^2 = \sup_{\substack{B \in Z \\ B \subset A}} \| G(B) S \|_2^2.$$

In particular, for $S = (\cdot, f) g$ we obtain

$$\| G(A) S \|_2^2 = \inf_{\substack{B \in Z \\ B \supset A}} \rho_{f,g}(B) = \sup_{\substack{B \in Z \\ B \subset A}} \rho_{f,g}(B) = \rho_{f,g}(A).$$

Let us now note that if we have $G(A) \neq 0$, then we can find an operator $S = (\cdot, f) g$ such that we have $G(A) S \neq 0$. But then we have $\rho_{f,g}(A) > 0$.

3) Let f and g be chosen such that we have $\rho_{f,g}(A) > 0$. Let us also fix ν in the range $0 < \nu < 1$. According to step 2), we can find a set B of the form

$$B = \bigcup_{i=1}^{\infty} \Delta_i \quad (\Delta_i = \delta_i \times \partial_i;\ \Delta_i \cap \Delta_j = \varnothing \text{ with } i \neq j)$$

such that we have

$$B \supset A \text{ and } \rho_{f,g}(B \setminus A) < \nu\rho_{f,g}(A).$$

For a rectangle Δ_{i_0} we have

$$\rho_{f,g}(\Delta_{i_0} \setminus A \cap \Delta_{i_0}) < \nu\rho_{f,g}(A \cap \Delta_{i_0}).$$

Let us denote this rectangle by Δ and the intersection $A \cap \Delta_{t_0}$ by A_0. We can now write

$$\rho_{f,g}(\Delta \setminus A_0) < \nu\rho_{f,g}(A_0). \tag{3.1}$$

Let us take $\Delta = \delta \times \partial$. Let us set*

$$f' = \frac{E(\delta) f}{\|E(\delta) f\|}, \quad g' = \frac{F(\partial) g}{\|F(\partial) g\|}.$$

This normalization maintains inequality (3.1), i.e., we have

$$\rho_{f',g'}(\Delta \setminus A_0) < \nu\rho_{f',g'}(A_0). \tag{3.2}$$

In the following we will simply write ρ instead of $\rho_{f',g'}$.

4) Let us now obtain the lower bound to the integral operator K corresponding to the elements f' and g'. Instead of K, let us consider the operator K_1 with the kernel

$$\varphi_1(\lambda, \mu) = e^{-\frac{2\Pi k_0}{n} t} \varphi(\lambda, \mu).$$

It is obvious that we have $\|K\| = \|K_1\|$. The following inequality holds for all (λ, μ) belonging to A_0:

$$\operatorname{Re} \varphi_1(\lambda, \mu) \geqslant M_\varepsilon \cos \frac{\Pi}{n}. \tag{3.3}$$

Let us consider the section of the set A_0 given by

$$A_0^\mu = \{\lambda \in \delta: (\lambda, \mu) \in A_0\} \quad (\mu \in \partial).$$

Let us take $0 < \varkappa < 1$. Let us consider the set

$$\partial_\varkappa = \{\mu \in \partial: \sigma_{f'}(A_0^\mu) > \varkappa\}.$$

It is easy to show that†

$$\tau_{g'}(\partial_\varkappa) \geqslant 1 - \frac{\nu}{1-\varkappa}.$$

If we set $\varkappa = 1 - \sqrt{\nu}$, we will find that

$$\tau_{g'}(\partial_\varkappa) \geqslant 1 - \sqrt{\nu} = \varkappa. \tag{3.4}$$

Let us now consider the function $u(\lambda) = \chi_\delta(\lambda)$ (the characteristic function of set δ). It is obvious that we have $\|u\|_{L_2(\sigma_{f'})} = 1$. Let us consider the function $v = K_1 u$, i.e.,

$$v(\mu) = \int_\delta \varphi_1(\lambda, \mu) \sigma_{f'}(d\lambda) \quad (\mu \in \partial).$$

For all $\mu \in \partial_\varkappa$ we have [because of (3.3)]

---

* Because of (3.1), we have $0 < \rho_{f,g}(A_0) \leqslant \rho_{f,g}(\Delta) = \|E(\delta) f\|^2 \|F(\partial) g\|^2$, and, consequently, $E(\delta)f$ and $F(\partial)g$ are both nonzero.

† Indeed, let us take $\Delta_1 = \delta \times \partial_\varkappa$, $\Delta_2 = \Delta \setminus \Delta_1$, $A_{0,1} = A_0 \cap \Delta_1$, and $A_{0,2} = A_0 \cap \Delta_2$. We then have $\tau_{g'}(\partial \setminus \partial_\varkappa) = \rho(\Delta_2) = \rho(\Delta) - \rho(\Delta_1) \leqslant (1+\nu) \times \rho(A_0) - \rho(A_{0,1}) = \rho(A_{0,2}) + \nu\rho(A_0) \leqslant \varkappa \cdot \tau_{g'}(\partial \setminus \partial_\varkappa) + \nu$. This yields $1 - \tau_{g'}(\partial_\varkappa) = \tau_{g'}(\partial \setminus \partial_\varkappa) \leqslant \frac{\nu}{1-\varkappa}$.

$$|v(\mu)| \geqslant \left| \int_{A_0^\mu} \varphi_1(\lambda, \mu) \sigma_{f'}(d\lambda) \right| - \left| \int_{\delta \setminus A_0^\mu} \varphi_1(\lambda, \mu) \sigma_{f'}(d\lambda) \right| \geqslant \varkappa M_\varepsilon \cos \frac{\Pi}{n} - (1-\varkappa) M.$$

We assume that $\varkappa$ is sufficiently close to unity (ν sufficiently close to zero), so that the right-hand side of this inequality is positive. It follows from this inequality and (3.4) that

$$\| v \|_{L_2(\tau_{g'})} \geqslant \left(\varkappa M_\varepsilon \cos \frac{\Pi}{n} - (1-\varkappa) M \right) \sqrt{\varkappa}.$$

We then have $m \geqslant \| K \| \geqslant \| v \|_{L_2(\tau_{g'})}$.

5) Hence, for (arbitrary) fixed $\varepsilon > 0$ ($\varepsilon < M$), $n > 2$, and all $\varkappa < 1$ sufficiently close to unity, we have the inequality

$$m \geqslant \left(\varkappa M_\varepsilon \cos \frac{\Pi}{n} - (1-\varkappa) M \right) \sqrt{\varkappa}.$$

The consecutive transitions to the limits

$$\varkappa \to 1, \quad n \to \infty, \quad \varepsilon \to 0$$

yield the inequality $m \geq M$.

The lemma has been proved.

THEOREM 3.1. If $\Phi \in (\mathbf{S}_{\mathrm{I}}, \mathbf{S}_{\mathrm{II}})$ then we have

$$\| \Phi \|_{\mathbf{S}_{\mathrm{I}}, \mathbf{S}_{\mathrm{II}}} \geqslant \| \Phi \|_{\mathbf{S}_2, \mathbf{S}_2} = G - \sup | \varphi(\lambda, \mu) |.$$

The proof of this assertion follows directly from Lemma 3.1 if it is recalled that the operator norm does not exceed the norm in s.n. ideals, while for one-dimensional operators the two norms are equal.

Corollary 3.1. The spaces $(\mathbf{S}_{\mathrm{I}}, \mathbf{S}_{\mathrm{II}})$ are complete.

Proof. Let us take $\Phi_n \in (\mathbf{S}_{\mathrm{I}}, \mathbf{S}_{\mathrm{II}})$ and

$$\| \Phi_n - \Phi_m \|_{\mathbf{S}_{\mathrm{I}}, \mathbf{S}_{\mathrm{II}}} \xrightarrow[n, m \to \infty]{} 0.$$

Since the space $[\mathbf{S}_{\mathrm{I}}, \mathbf{S}_{\mathrm{II}}]$ is complete, the sequence Φ_n in it possesses the limit Φ' given by

$$\| \Phi_n - \Phi' \|_{\mathbf{S}_{\mathrm{I}}, \mathbf{S}_{\mathrm{II}}} \to 0, \quad \Phi' \in [\mathbf{S}_{\mathrm{I}}, \mathbf{S}_{\mathrm{II}}].$$

Moreover, in view of Theorem 3.1 and the completeness of the ring $r(\mathbf{S}_2)$, the sequence Φ_n possesses a limit Φ in $r(\mathbf{S}_2)$

$$\| \Phi_n - \Phi \|_{\mathbf{S}_2, \mathbf{S}_2} \to 0, \quad \Phi \in r(\mathbf{S}_2).$$

Let us take an arbitrary operator $S \in \mathbf{S}_{\mathrm{I}} \cap \mathbf{S}_2$. We then have

$$\| \Phi_n S - \Phi' S \|_{\mathbf{S}_{\mathrm{II}}} \to 0, \quad \| \Phi_n S - \Phi S \|_{\mathbf{S}_2} \to 0,$$

and, consequently,

$$\Phi' S = \Phi S \quad (S \in \mathbf{S}_{\mathrm{I}} \cap \mathbf{S}_2).$$

It follows from this that we have

$$\Phi \in (\mathbf{S}_{\mathrm{I}}, \mathbf{S}_{\mathrm{II}}) \text{ and } \Phi = \Phi'.$$

§ 4. The Rings m(S)

In this section we consider the rings of transformers in arbitrary s.n. ideals **S** of the ring **R** (we do not exclude the case **S** = **R**). A ring m(**S**) of transformers can be defined in every s.n. ideal **S**. When the ideal **S** is separable or we have $\mathbf{S} \subset \mathbf{S}_2$, the rings m(**S**) are subrings of the rings r(**S**). We establish some of the properties of these rings; in particular, we describe the space of maximal ideals. If the ideal S is contained in $\mathbf{S}_2$, or is separable, or is conjugate to a separable s.n. ideal, then the existence of a natural involution and the absence of the radical in the rings m(**S**) is proved.

We will now assume that Λ and M are Hausdorff spaces with denumerable bases. Concerning the spectral measures $E(\cdot)$ and $F(\cdot)$ we will assume that all Borel sets (belonging to Λ and M) are measurable. It is easy to see that in this case all of the Borel sets of the Hausdorff space with base $\Lambda \times$ M will also be measurable in the spectral measure G.

By the carrier of a spectral measure defined on a topological space we mean the set of all points of this space, each open neighborhood of which has a nonzero spectral measure. The carrier of a measure is a closed set. If the space has a denumerable basis, then the complement of the measure carrier has a spectral measure zero.

The carriers of $E(\cdot)$ and $F(\cdot)$ are denoted by Λ_0 and M_0. Let us show that the carrier of measure G (denoted by supp G) is the product $\Lambda_0 \times M_0$. Indeed, if $(\lambda, \mu) \overline{\in} \Lambda_0 \times M_0$, then, for example, we have $\lambda \overline{\in} \Lambda_0$. Consequently, the point λ has a neighborhood $U(\lambda)$ of E-measure zero. But the point (λ, μ) will then have a neighborhood $U(\lambda) \times M$ of G-measure zero and, therefore, it is not contained in supp G. Hence, supp $G \subset \Lambda_0 \times M_0$. Conversely, let us assume that $(\lambda, \mu) \in \Lambda_0 \times M_0$. Any neighborhood of the point (λ, μ) contains a neighborhood of the form $\Delta = \delta \times \partial$. Since $\lambda \in \Lambda_0$ and $\mu \in M_0$, we have $E(\delta) \neq 0$ and $F(\partial) \neq 0$. However, now the transformer $G(\Delta)$ defined by

$$G(\Delta) S = F(\partial) SE(\delta) \quad (S \in \mathbf{S}_2)$$

is also nonzero. We therefore have $\Lambda_0 \times M_0 \subset \operatorname{supp} G$. We will assume in the following that the measures $E(\cdot)$, $F(\cdot)$, and, consequently, $G(\cdot)$ have the compact carriers Λ_0, M_0, and $\Lambda_0 \times M_0$, respectively.

Let us consider the set $F \subset C(\Lambda_0 \times M_0)$ of all complex-valued functions of the form

$$\varphi(\lambda, \mu) = \sum_{i=1}^{n} x_i(\lambda) y_i(\mu),$$

where

$$x_i \in C(\Lambda_0), \quad y_i \in C(M_0) \quad (i = 1, \ldots, n).$$

F is an open subring of the ring $C(\Lambda_0 \times M_0)$ and is dense in the latter. Let **S** be a s.n. ideal of the ring **R** (the case **S** = **R** is not excluded). Every function $\varphi \in F$ generates a bounded linear transformer $\Phi_{\mathbf{S}}$ in **S** according to the following rule: If

$$\varphi(\lambda, \mu) = \sum_{i=1}^{n} x_i(\lambda) y_i(\mu),$$

then we have

$$\Phi_{\mathbf{S}} S = \sum_{i=1}^{n} Y_i S X_i \quad (S \in \mathbf{S}), \tag{4.1}$$

where

$$X_i = \int_\Lambda x_i(\lambda) E(d\lambda), \quad Y_i = \int_M y_i(\mu) F(d\mu) \quad (i = 1, \ldots, n).$$

The mapping

$$\varphi \to \Phi_S \quad (\varphi \in F) \tag{4.2}$$

is a homomorphism of ring F in to the ring [S, S]. The linearity of the mapping is obvious. In order to show that a product of functions maps into a product of transformers, we have only to consider functions of the form

$$\varphi_1(\lambda, \mu) = x_1(\lambda) y_1(\mu) \text{ and } \varphi_2(\lambda, \mu) = x_2(\lambda) y_2(\mu);$$

our assertion becomes obvious for these functions when we note that we have

$$\int_\Lambda x_1(\lambda) x_2(\lambda) E(d\lambda) = \int_\Lambda x_1(\lambda) E(d\lambda) \int_\Lambda x_2(\lambda) E(d\lambda),$$

$$\int_M y_1(\mu) y_2(\mu) F(d\mu) = \int_M y_1(\mu) F(d\mu) \int_M y_2(\mu) F(d\mu).$$

The image of the ring F under the homomorphism (4.2) is denoted by R(S).

Remark 4.1. The transformer Φ_S generated by the function $\varphi \in F$ is the limit in [S, S] of integral sums of the form

$$\sum_{l=1}^{n} \sum_{m=1}^{p} \varphi(\lambda_l, \mu_m) F(\partial_m)(\cdot) E(\delta_l).$$

Here, $\{\delta_l\}_1^n$ and $\{\partial_m\}_1^p$ are partitionings of the sets Λ_0 and M_0, respectively, and $\lambda_l \in \delta_l$, $\mu_m \in \partial_m$. The limit is taken under the condition that $\max_{l,m}(\operatorname{diam} \delta_l, \operatorname{diam} \delta_m) \to 0$. *

It is sufficient to prove this assertion for functions of the form: $\varphi(\lambda, \mu) = x(\lambda)y(\mu)$. But in this case, the integral sum is of the form

$$\left[\sum_{m=1}^{p} y(\mu_m) F(\partial_m)\right](\cdot)\left[\sum_{l=1}^{n} x(\lambda_l) E(\delta_l)\right].$$

It remains to note that in the limit we have

$$\sum_{l=1}^{n} x(\lambda_l) E(\delta_l) \to \int_{\Lambda_0} x(\lambda) E(d\lambda),$$

$$\sum_{m=1}^{p} y(\mu_m) F(\partial_m) \to \int_{M_0} y(\mu) F(d\mu)$$

in the norm in **R**.

It follows from this proof that Φ_S is independent of the representation of the function $\varphi \in F$, but is governed by the function φ itself.

---

*The spaces Λ_0 and M_0 are metrizable as compact Hausdorff spaces with denumerable bases.

Remark 4.2. With the help of Remark 4.1 we can establish the connection between transformers defined by formula (4.1) and transformers defined by double-integral operators of the form of (2.2). It follows directly from Remark 4.1 that when $\mathbf{S} = \mathbf{S}_2$ the transformer coincides with the integral (2.2), i.e., we have

$$\Phi_{\mathbf{S}_2} = \int_{\Lambda_0} \int_{M_0} \varphi(\lambda, \mu)\, dG(e) \qquad (\varphi \in F).$$

When $\mathbf{S}$ is an arbitrary s.n. ideal, we have for all operators $S \in \mathbf{S} \cap \mathbf{S}_2$

$$\Phi_{\mathbf{S}} S = \Phi_{\mathbf{S}_2} S = \int_{\Lambda_0} \int_{M_0} \varphi(\lambda, \mu)\, dG(e)\, S.$$

It follows from this that if the ideal $\mathbf{S}$ is separable or we have $\mathbf{S} \subset \mathbf{S}_2$, then the transformer $\Phi_{\mathbf{S}}$ coincides with the transformer $\Phi_{\mathbf{S},\mathbf{S}}$ generated by the same function $\varphi(\lambda, \mu)$ in the sense of Section 2.

Let us obtain the lower bound to the norm of $\Phi_{\mathbf{S}}$. If we take account of Remark 4.2, then Lemma 3.1 directly yields

$$\| \Phi_{\mathbf{S}} \|_{\mathbf{S},\mathbf{S}} \geqslant \| \varphi \|_{C(\Lambda_0 \times M_0)} \qquad (\varphi \in F). \tag{4.3}$$

The homomorphism (4.2) therefore has an inverse

$$\Phi_{\mathbf{S}} \to \varphi \qquad (\Phi_{\mathbf{S}} \in R(\mathbf{S})), \tag{4.4}$$

continuous as an operator from $[\mathbf{S}, \mathbf{S}]$ into $C(\Lambda_0 \times M_0)$. The closure of the ring $R(\mathbf{S})$ in $[\mathbf{S}, \mathbf{S}]$ will be denoted by $m(\mathbf{S})$. The homomorphism (4.4) can be extended to $m(\mathbf{S})$.

Remark 4.3. All that was stated in Remark 4.2 about the transformers $\Phi_{\mathbf{S}} \in R(\mathbf{S})$ can be transferred to any transformers $\Phi_{\mathbf{S}} \in m(\mathbf{S})$. The function $\varphi(\lambda, \mu)$ should always be understood to be the image of the transformer $\Phi_{\mathbf{S}}$ under the homomorphism (4.4).

The ring $m(\mathbf{S}_2)$ is obviously the totality of all double-integral operators

$$\Phi_{\mathbf{S}_2} = \int \int \varphi(\lambda, \mu)\, dG(e),$$

where φ is an arbitrary function belonging to $C(\Lambda_0 \times M_0)$. The ring $m(\mathbf{S}_2)$ is isometrically isomorphic to the algebra $C(\Lambda_0 \times M_0)$.

If the ideal $\mathbf{S}$ is separable or we have $\mathbf{S} \subset \mathbf{S}_2$, then the ring $m(\mathbf{S})$ is a closed subring of the ring $r(\mathbf{S})$. It should also be noted that the totality of all transformers belonging to $r(\mathbf{S})$ which are generated by continuous functions also forms a closed subring $m'(\mathbf{S})$ of the ring $r(\mathbf{S})$. It is obvious that we have

$$m(\mathbf{S}) \subset m'(\mathbf{S}).$$

Generally speaking, however, the rings $m(\mathbf{S})$ and $m'(\mathbf{S})$ need not coincide when $\mathbf{S} \neq \mathbf{S}_2$.

Remark 4.4. Let $\mathbf{S}_{\mathrm{I}}$ and $\mathbf{S}_{\mathrm{II}}$ be two s.n. ideals, where $\mathbf{S}_{\mathrm{I}} \subset \mathbf{S}_{\mathrm{II}}$ and the norm in $\mathbf{S}_{\mathrm{I}}$ is induced by the norm in $\mathbf{S}_{\mathrm{II}}$. If the transformer $\Phi_{\mathbf{S}_{\mathrm{II}}}$ belongs to $m(\mathbf{S}_{\mathrm{II}})$, then its restriction $\Phi_{\mathbf{S}_{\mathrm{II}}}|_{\mathbf{S}_{\mathrm{I}}}$ belongs to $m(\mathbf{S}_{\mathrm{I}})$, while the transformers $\Phi_{\mathbf{S}_{\mathrm{II}}}$ and $\Phi_{\mathbf{S}_{\mathrm{II}}}|_{\mathbf{S}_{\mathrm{I}}}$ have the same image $\varphi(\lambda, \mu)$ under the homomorphism (4.4).

Indeed, let us assume that

$$\Phi_{S_{II}}^{(n)} \in R(S_{II}), \quad \left\| \Phi_{S_{II}}^{(n)} - \Phi_{S_{II}} \right\|_{S_{II}, S_{II}} \to 0.$$

Let us assume that under the homomorphism (4.4) the transformers $\Phi_{S_{II}}^{(n)}$ are mapped into the functions $\varphi_n \in F$, and the transformer $\Phi_{S_{II}}$ is mapped into the function $\varphi \in C(\Lambda_0 \times M_0)$. We then have

$$\| \varphi_n - \varphi \|_C \to 0.$$

The functions φ_n generate the transformers

$$\Phi_{S_I}^{(n)} = \Phi_{S_{II}}^{(n)} \Big|_{S_I} \in R(S_I).$$

It is obvious that we have

$$\left\| \Phi_{S_{II}}^{(n)} \Big|_{S_I} - \Phi_{S_{II}} \Big|_{S_I} \right\|_{S_I, S_I} \to 0.$$

The remark has been proved.

Let us now find the maximal ideals of the rings m(S). Let us first of all consider transformers that are induced by the functions

$$\varphi(\lambda, \mu) = x(\lambda) \quad (x \in C(\Lambda_0)).$$

Transformer Φ_S acts according to the formula

$$\Phi_S S = SX \quad (S \in S), \tag{4.5}$$

where $X = \int_{\Lambda_0} x(\lambda) E(d\lambda)$. As is well known, we have

$$\| \Phi_S \|_{S, S} = \| X \| = \| x \|_{C(\Lambda_0)}.$$

The subring $m_X(S)$ of the ring m(S) formed from all transformers of the form of (4.5) is isometrically isomorphic to the ring $C(\Lambda_0)$. Any nontrivial homomorphism of this ring in the complex plane has the form

$$\Phi_S \to x(\lambda^*),$$

where

$$\lambda^* \in \Lambda_0. \tag{4.6}$$

In the analogous manner we find that the ring $m_Y(S)$ of all transformers Φ_S generated by functions of the form $\varphi(\lambda, \mu) = y(\mu)$, with $y \in C(M_0)$, is isometrically isomorphic to the ring $C(M_0)$. The general form of the nonzero multiplicative functional on this ring is given by the formula

$$\Phi_S \to y(\mu^*),$$

where

$$\mu^* \in M_0. \tag{4.7}$$

Let f_0 be a nontrivial multiplicative function defined on the ring m(S). The restrictions of this functional to the subrings $m_X(S)$ and $m_Y(S)$ are nonzero multiplicative linear functionals and, therefore, of the form of (4.6) and (4.7). For all transformers Φ_S generated by functions

$$\varphi(\lambda, \mu) = \sum_{i=1}^{n} x_i(\lambda) y_i(\mu),$$

we have

$$f_0(\Phi_S) = \varphi(\lambda^*, \mu^*) \quad (\Phi_S \in R(S)).$$

On the other hand, since the mapping (4.4) is a nontrivial continuous homomorphism of the ring m(S) into the ring $C(\Lambda_0 \times M_0)$, the functional f

$$f(\Phi_S) = \varphi(\lambda^*, \mu^*) \quad (\Phi_S \in m(S))$$

is a nonzero linear multiplicative functional in the ring m(S) for all points $(\lambda^*, \mu^*) \in \Lambda_0 \times M_0$.

As the set R(S) is dense in m(S), the functional f_0 coincides with f.

We have thus proved the following assertion.

THEOREM 4.1. *The general form of a nontrivial multiplicative functional over the ring m(S) is given by the formula*

$$f(\Phi_S) = \varphi(\lambda^*, \mu^*) \quad ((\lambda^*, \mu^*) \in \Lambda_0 \times M_0)$$

[the function φ is the image of the transformer Φ_S under the homomorphism (4.4)].

The space of the maximal ideals of the ring m(S) is homeomorphic to the space $\Lambda_0 \times M_0$.

Remark 4.5. We were able to find the space of the maximal ideals of the rings m(S) so simply because we constructed the rings m(S) in an appropriate manner. The maximal ideals of the rings m'(S) defined in Remark 4.4 are unknown to the author.

We will assume in the following that the ideal S is contained in S_2, or is separable, or is conjugate to a separable s.n. ideal $\tilde{S}$, i.e., $S = \tilde{S}^*$.

It should be noted that an analog of Lemma 2.1 holds for the rings m(S), where the ideal S is either separable or is contained in S_2.

In these rings there exists a symmetric involution $\Phi_S \to \overline{\Phi}_S$ [i.e., the homomorphism (4.4) maps the transformers Φ_S and $\overline{\Phi}_S$ into a pair of complex conjugate functions] and in addition we have

$$\|\Phi_S\|_{S,S} = \|\overline{\Phi}_S\|_{S,S}. \quad (4.8)$$

Indeed, since we have $m(S) \subset r(S)$, it is sufficient to show that the involution $\Phi \to \overline{\Phi}$, defined in the ring r(S) by Lemma 2.1 maps the ring m(S) into itself. Moreover, since the involution $\Phi \to \overline{\Phi}$ is continuous in r(S), it is sufficient to show that it maps the set R(S) dense in m(S) into the ring m(S). The latter assertion, however, is obvious because the transformers $\Phi_S \in R(S)$ are generated by the functions $\varphi(\lambda, \mu) \in F$, while the set F contains both $\varphi(\lambda, \mu)$ and $\varphi(\lambda, \mu)$.

Remark 4.6. If S is an arbitrary s.n. ideal, then the ring R(S) also contains the involution $\Phi \to \overline{\Phi}$. Unfortunately, the author has not been able to establish the continuity of this involution in the general case and hence to obtain its closure to the involution of the ring m(S).

The connection between the rings $m(\tilde{S})$ and $m(S)$ (where $S = \tilde{S}^*$) is provided by the following lemma.

Lemma 4.1. *The rings $m(\tilde{S})$ and $m(S)$ are isometrically isomorphic to one another. The isomorphism is given by the formula*

$$\Phi_{\tilde{S}} \to (\overline{\Phi}_{\tilde{S}})^* = \Phi_S. \quad (4.9)$$

The transformers $\Phi_{\tilde{S}}$ and Φ_S have the same image $\varphi(\lambda, \mu)$ under the homomorphism (4.4).

Proof. The mapping (4.9) is an isometric isomorphism of the ring $m(\tilde{S})$ onto a closed subring $m_0(S)$ of the ring $[S, S]$. We will show that $m_0(S) = m(S)$.

First of all, let us note that the transformer conjugate to the transformer of multiplication on the left (right) by the operator $A \in R$ in ideal $\tilde{S}$ is the transformer of multiplication on the left (right) by the conjugate operator A^* in ideal S. Indeed, we have

$$\mathrm{sp}\,(AS)\,T^* = \mathrm{sp}\,ST^*A = \mathrm{sp}\,S\,(A^*T)^*,$$
$$\mathrm{sp}\,(SA)\,T^* = \mathrm{sp}\,S\,(TA^*)^* \quad (S \in \tilde{S},\ T \in S).$$

It follows from this that in the case of functions $\varphi \in F$ the transformer $(\overline{\Phi}_{\tilde{S}})^*$ coincides with the transformer Φ_S induced by the function $\varphi(\lambda, \mu)$ in ideal S according to formula (4.1). Hence, the isometric isomorphism (4.9) maps the dense set $R(\tilde{S}) \subset m(\tilde{S})$ onto the dense set $R(S) \subset m(S)$, the transformers

$$\Phi_{\tilde{S}} \text{ and } (\overline{\Phi}_{\tilde{S}})^* \quad (\Phi_{\tilde{S}} \in R(\tilde{S}))$$

having the same function $\varphi(\lambda, \mu)$ as image. This leads to the equality $m_0(S) = m(S)$, and the second assertion of the lemma.

Remark 4.7. The isomorphism between the rings $m(\tilde{S})$ and $m(S)$ (where $S = \tilde{S}^*$) generates in the ring $m(S)$ a symmetric involution $\Phi_S \to \overline{\Phi}_S$, with

$$\|\overline{\Phi}_S\|_{S,S} = \|\Phi_S\|_{S,S}.$$

Let us now turn to the question of the invertibility of the homomorphism (4.4), i.e., the question of whether ring $m(S)$ is without radical. If the homomorphism (4.4) is only considered on the ring $R(S)$, then it is invertible. In order for it to remain invertible after closure, it is necessary and sufficient that the topologies in the rings $R(S)$ and $C(\Lambda_0 \times M_0)$ be matched in the following manner. If transformers $\Phi_S^{(n)} \in R(S)$ form a fundamental sequence in $R(S)$ and we have $\|\varphi_n\|_{C(\Lambda_0 \times M_0)} \to 0$, then we must also have

$$\|\Phi_S^{(n)}\|_{S,S} \to 0.$$

The author has been unable to establish the invertibility of the homomorphism (4.4) for arbitrary s.n. ideals S. However, if we restrict ourselves to ideals of the three types indicated below, we can easily obtain the following result.

THEOREM 4.2. $m(S)$ is a ring without radical.

Proof. a) Let the ideal S be either separable or contained in S_2. If $\Phi_S \in m(S)$ is mapped into the function $\varphi(\lambda, \mu)$ under the homomorphism (4.4), then by Remark 4.3 the transformer Φ_S is generated by this function in the sense of the definition given in Section 2. Consequently, if $\varphi(\lambda, \mu) \equiv 0$, then we have $\Phi_S = 0$. Thus, the ring $m(S)$ is without radical.

b) Let the ideal S be conjugate to a separable ideal $\tilde{S}$, so that $S = \tilde{S}^*$. The assertion of the theorem then follows from point a) and Lemma 4.1.

The theorem has been proved.

Let us now consider the following four s.n. ideals:

$$S_I = S_\Psi, \quad S_{II} = S_\Psi^{(0)}, \quad S_{III} = S_{\Psi^*}, \quad S_{IV} = S_{\Psi^*}^{(0)}.$$

Here, Ψ is a symmetric normalizing function (the definitions of these ideals are given in Chapter 3 of [3]).

Let us introduce the following abbreviations:

$$\Phi_{S_i} = \Phi_i, \quad \|\Phi_{S_i}\|_{S_i, S_i} = \|\Phi_i\|_{i, i}, \quad m(S_i) = m_i \quad (i = \mathrm{I, II, III, IV}).$$

If the function $\varphi(\lambda, \mu)$ generates the transformer $\Phi_{\mathrm{I}}(\Phi_{\mathrm{III}})$ of class $m_{\mathrm{I}}(m_{\mathrm{III}})$, † then by Remark 4.4 it also generates the transformer $\Phi_{\mathrm{II}}(\Phi_{\mathrm{IV}})$ of class $m_{\mathrm{II}}(m_{\mathrm{IV}})$ and

$$\Phi_{\mathrm{II}} = \Phi_{\mathrm{I}}|_{S_{\mathrm{II}}} \quad \left(\Phi_{\mathrm{IV}} = \Phi_{\mathrm{III}}|_{S_{\mathrm{IV}}}\right).$$

In fact, the following holds: If the function $\varphi(\lambda, \mu)$ generates a transformer of one of the classes m_{I}, m_{II}, m_{III}, or m_{IV}, then it also generates transformers of the remaining three classes and, in addition, the norms of these transformers are equal

$$\|\Phi_{\mathrm{I}}\|_{\mathrm{I, I}} = \|\Phi_{\mathrm{II}}\|_{\mathrm{II, II}} = \|\Phi_{\mathrm{III}}\|_{\mathrm{III, III}} = \|\Phi_{\mathrm{IV}}\|_{\mathrm{IV, IV}}.$$

More accurately, we have the following theorem.

THEOREM 4.3. *The rings m_{I}, m_{II}, m_{III}, and m_{IV} are isometrically isomorphic. The isomorphisms between them can be established by either of the two equivalent methods:*

a) By means of the diagram

$$\begin{array}{ccc} m_{\mathrm{I}} \ni (\bar{\Phi}_{\mathrm{IV}})^* = \Phi_{\mathrm{I}} & \leftrightarrow & \Phi_{\mathrm{I}}|_{S_{\mathrm{II}}} = \Phi_{\mathrm{II}} \in m_{\mathrm{II}}, \\ \updownarrow & & \updownarrow \\ m_{\mathrm{IV}} \ni \Phi_{\mathrm{III}}|_{S_{\mathrm{IV}}} = \Phi_{\mathrm{IV}} & \leftrightarrow & (\bar{\Phi}_{\mathrm{II}})^* = \Phi_{\mathrm{III}} \in m_{\mathrm{III}}. \end{array}$$

b) The isomorphisms are established so as to make the same function $\varphi(\lambda, \mu)$ correspond to all of the transformers.

Proof. First of all, let us note that S_{II} is a subideal of ideal S_{I}, $S_{\mathrm{III}} = S_{\mathrm{II}}^*$, S_{IV} is a subideal of ideal S_{III}, and $S_{\mathrm{I}} = S_{\mathrm{IV}}^*$. Therefore, if we move clockwise in the above diagram, then on each step the corresponding mappings

$$\Phi_{\mathrm{I}} \to \Phi_{\mathrm{II}} \ (m_{\mathrm{I}} \text{ into } m_{\mathrm{II}}), \quad \Phi_{\mathrm{II}} \to \Phi_{\mathrm{III}} \ (m_{\mathrm{II}} \text{ into } m_{\mathrm{III}}),$$
$$\Phi_{\mathrm{III}} \to \Phi_{\mathrm{IV}} \ (m_{\mathrm{III}} \text{ into } m_{\mathrm{IV}}), \quad \Phi_{\mathrm{IV}} \to \Phi_{\mathrm{I}} \ (m_{\mathrm{IV}} \text{ into } m_{\mathrm{I}})$$

are homomorphisms; in every case, the image and inverse image are generated by the same function $\varphi(\lambda, \mu)$ (see Remark 4.4 and Lemma 4.1). Moreover, the norm of the image does not exceed that of the inverse image,

$$\|\Phi_{\mathrm{I}}\|_{\mathrm{I, I}} \geqslant \|\Phi_{\mathrm{II}}\|_{\mathrm{II, II}} = \|\Phi_{\mathrm{III}}\|_{\mathrm{III, III}} \geqslant \|\Phi_{\mathrm{IV}}\|_{\mathrm{IV, IV}} = \|\Phi_{\mathrm{I}}\|_{\mathrm{I, I}}.$$

Since the diagram is closed, it does indeed define an isometric isomorphism of the corresponding rings at each step and, in addition, in a manner such that corresponding transformers are generated by the same function $\varphi(\lambda, \mu)$.

† This means that the transformer $\Phi_{\mathrm{I}}(\Phi_{\mathrm{III}})$ and the function $\varphi(\lambda, \mu)$ are associated with one another by the homeomorphism (4.4).

Finally let us extend Lemma 2.2 to the case under consideration, i.e., let us prove that the transformers Φ_S are coordinated in the following sense.

LEMMA 4.2. Let $\mathbf{S}_I$ and $\mathbf{S}_{II}$ be two s.n. ideals of the three types indicated above. If the transformers $\Phi_I \in m_I$ and $\Phi_{II} \in m_{II}$ correspond to the same function $\varphi(\lambda, \mu)$, then for all $S \in \mathbf{S}_I \cap \mathbf{S}_{II}$ we have

$$\Phi_I S = \Phi_{II} S.$$

Proof. Let us consider a number of cases:

a) Either one of the ideals $\mathbf{S}_I$ and $\mathbf{S}_{II}$ is contained in $\mathbf{S}_2$ or both ideals $\mathbf{S}_I$ and $\mathbf{S}_{II}$ are separable. If we take Remark 4.3 into account, then it is clear that Lemma 4.2 in this case is identical with Lemma 2.2;

b) $\mathbf{S}_I$ and $\mathbf{S}_{II}$ are conjugate to the separable ideals $\tilde{\mathbf{S}}_I$ and $\tilde{\mathbf{S}}_{II}$, i.e., $\mathbf{S}_I = \tilde{\mathbf{S}}_I^*$ and $\mathbf{S}_{II} = \tilde{\mathbf{S}}_{II}^*$. For any operators

$$S \in \mathbf{S}_I \cap \mathbf{S}_{II}, \quad T \in \tilde{\mathbf{S}}_I \cap \tilde{\mathbf{S}}_{II}$$

we have the equalities [by Lemma 4.1 and item a)]

$$\operatorname{sp} T(\Phi_I S)^* = \operatorname{sp}\left(\bar{\Phi}_{\tilde{\mathbf{S}}_I} T\right) S^* = \operatorname{sp}\left(\bar{\Phi}_{\tilde{\mathbf{S}}_{II}} T\right) S^* = \operatorname{sp} T(\Phi_{II} S)^*.$$

Consequently, we have $\Phi_I S = \Phi_{II} S$;

c) $\mathbf{S}_I$ is separable and $\mathbf{S}_{II}$ is conjugate to a separable ideal.

We then have $\mathbf{S}_I = \mathbf{S}_\Psi^{(0)} \subset \mathbf{S}_\Psi$ (where Ψ is a s.n. function). According to Theorem 4.3, the function $\varphi(\lambda, \mu)$ generates the transformer $\Phi_{\mathbf{S}_\Psi}$ of class $m(\mathbf{S}_\Psi)$ and $\Phi_I = \Phi_{\mathbf{S}_\Psi}|_{\mathbf{S}_I}$. Such ideal $\mathbf{S}_\Psi$ is conjugated to the separable ideal $\mathbf{S}_{\Psi^*}^{(0)}$. According to point b), for all operators $S \in \mathbf{S}_{II} \cap \mathbf{S}_\Psi$ we have

$$\Phi_{II} S = \Phi_{\mathbf{S}_\Psi} S.$$

Consequently, for all $S \in \mathbf{S}_I \cap \mathbf{S}_{II}$ we have

$$\Phi_I S = \Phi_{\mathbf{S}_\Psi} S = \Phi_{II} S.$$

The lemma has been proved.

Remark 4.8. Let us recall that in Section 4 we have counted the whole of the ring $\mathbf{R}$ among the s.n. ideals of $\mathbf{R}$. In particular, because $\mathbf{R} = \mathbf{S}_1^*$, $\mathbf{S}_\infty = \mathbf{R}^{(0)}$, $\mathbf{S}_1 = \mathbf{S}_\infty^*$, $\mathbf{S}_1 = \mathbf{S}_1^{(0)}$, the rings $m(\mathbf{S}_1)$, $\mathbf{S}(\mathbf{R})$, and $m(\mathbf{S}_\infty)$ are isometrically isomorphic by Theorem 4.3. The analogous fact for the rings $r(\mathbf{S}_1)$, $r(\mathbf{R})$, and $r(\mathbf{S}_\infty)$ was noted in [2].

The author would like to express his gratitude to M. Z. Solomyak for his supervision and help with the work.

Literature Cited

1. M. Sh. Birman and M. Z. Solomyak, "Stieltjes double-integral operators," in: Topics in Mathematical Physics, Vol. 1, M. Sh. Birman (editor), Consultants Bureau, New York (1967), pp. 25-54.
2. M. Sh. Birman and M. Z. Solomyak, "Stieltjes double-integral operators. II," in: Topics in Mathematical Physics, Vol. 2, M. Sh. Birman (editor), Consultants Bureau, New York (1968), pp. 19-46.

3. I. Ts. Gokhberg and M. G. Krein, Introduction to the Theory of Linear Nonself-Adjoint Operators in Hilbert Space [in Russian], Izd. "Nauka," Moscow (1965).
4. I. M. Gel'fand, A. A. Raikov, and G. E. Shilov, Commutative Normed Rings [in Russian], Fizmatgiz, Moscow (1960).
5. Yu. M. Berezanskii, "Spaces with a negative norm," Uspekhi Matem. Nauk, Vol. 18, No. 1 (1963).
6. A. I. Plesner and V. A. Rokhlin, "Spectral theory of operators," Part II, Uspekhi Matem. Nauk, Vol. 1, No. 1(11) (1946).

ON MULTIDIMENSIONAL INTEGRAL OPERATORS

B. S. Pavlov

In a number of problems of operator theory, it is necessary to investigate integral operators of the form

$$\int\int \varphi(\lambda, \mu)\, dE_\lambda T\, dE_\mu.$$

Such integral operators have been studied in a number of articles by M. Sh. Birman and M. Z. Solomyak ([1], [2], [3]).

The present article is devoted to an investigation of some of the properties of the analogous multiple integrals

$$\int\int \dots \int \varphi(\lambda_1, \lambda_2, \dots, \lambda_n)\, dE^1_{\lambda_1} T_1\, dE^2_{\lambda_2} T_2 \dots T_{n-1}\, dE^n_{\lambda_n}. \tag{1}$$

Such integrals arise, for example, in the construction of perturbation-theory series in quantum mechanics.

The first section of this article is devoted to the case when the operators $T_1, T_2, \dots, T_n$ are Hilbert–Schmidt operators. It is found that the operator function of the multidimensional intervals $\Delta = \Delta_1 \times \Delta_2 \times \dots \times \Delta_n$

$$m(\Delta) = E^1(\Delta_1)\, T_1 E^2(\Delta_2)\, T_2 \dots T_{n-1} E^n(\Delta_n) \tag{2}$$

is denumerably additive in $\mathfrak{S}_2$ and has a weakly bounded variation. This allows us to integrate any measurable bounded function with respect to it.

Multiple integral operators are investigated in the second part of the particle under the condition that the T_i are bounded operators. It should be noted that now the function (2) no longer has a bounded variation. However, in the same way as in [1] and [2], it is possible to give a meaning to integral (1) on the assumption that the integrand is sufficiently smooth. The requirement that the function $\varphi(\lambda_1, \lambda_2, \dots, \lambda_n)$ be smooth is expressed in terms of the boundedness of various "mixed" norms (see also [3, 4]).

It should be noted that in a number of special cases it is possible to ascribe a meaning to integral (1) under less stringent requirements on the smoothness of $\varphi(\lambda_1, \lambda_2, \dots, \lambda_n)$. In the important case when the integrand is a difference relation of order n − 1 in a function of one variable, this problem has been solved in M. Z. Solomyak and V. V. Sten'kin's article [5] included in the present volume.

We use the standard notation in the text: R is the class of bounded operators in Hilbert space H, $\mathbf{S}_2$ is the class of Hilbert–Schmidt operators, and $\mathbf{S}_1$ is the class of kernel operators. In the following, we also encounter "fractional" norms of functions of several variables defined by

$$\|u\|^2_{L_2^l} = \sum \int\limits_{\Omega\times\Omega}\int \frac{|D^{[l]}u(x) - D^{[l]}u(y)|^2}{|x-y|^{n+2(l-[l])}}\,dx\,dy. \tag{3}$$

Here, Ω is a domain in E_n, $[l]$ is the integral part of l (in our case, l is not an integer), $D^{[l]}u(x)$ is any partial derivative of order $[l]$, and the summation is extended over all partial derivatives of order $[l]$. The norm $\|u\|_{W_2^l}$ is defined by the equality

$$\|u\|^2_{W_2^l} = \sum |l_i(u)|^2 + \|u\|^2_{L_2^l}, \tag{4}$$

where $\{l_i\}$ is any complete system of linear functionals in the subspace of polynomials P(x) such that $\|P\|_{L_2^l} = 0$.

§ 1

Let (Ω, F, E) be a space Ω in which a spectral measure E acting in a separable Hilbert space H is defined on the σ-algebra F. Let us assume that we have several such spaces: $(\Omega^1, F^1, E^1), (\Omega^2, F^2, E^2), \ldots, (\Omega^n, F^n, E^n)$ and a collection of Hilbert–Schmidt operators T_1, $T_2, \ldots, T_n$. Let us consider the space $\Omega = \Omega^1 \times \Omega^2 \times \ldots \times \Omega^n$. In space Ω we define a set function

$$\Delta = \Delta_1 \times \Delta_2 \times \ldots \times \Delta_n, \quad \Delta_i \in F^i,$$

on the system of sets

$$m(\Delta) = E^1(\Delta_1)\,T_1 E^2(\Delta_2)\,T_2 \ldots T_{n-1}E^n(\Delta_n),$$

which in the following are called intervals. This function is additive. We will show that it has a weakly-bounded $\mathbf{S}_2$-variation and is denumerably additive. It is enough to show that for an arbitrary partitioning $\Lambda = \{\Delta_{i_1 i_2 \ldots i_n}\}$ of space Ω, i.e.,

$$\Omega = \sum_{i_1 i_2 \ldots i_n} \Delta_{i_1 i_2 \ldots i_n},$$

and for an arbitrary operator T_n belonging to $\mathbf{S}_2$ we have

$$\sum_{i_1 i_2 \ldots i_n} |\mathrm{Sp}\, m(\Delta_{i_1 i_2 \ldots i_n})\,T_n| \leqslant \|T_1\|_{\mathbf{S}_2}\|T_2\|_{\mathbf{S}_2} \ldots \|T_n\|_{\mathbf{S}_2}.$$

Let us note, first of all, that we have

$$\sum_{i_1 i_2 \ldots i_n} |\mathrm{Sp}\, m(\Delta_{i_1 i_2 \ldots i_n})\,T_n| \leqslant \sum_{i_1 i_2 \ldots i_n} \|E^1(\Delta_{i_1})\,T_1E^2(\Delta_{i_2})\|_{\mathbf{S}_2}\,\|E^2(\Delta_{i_2})\,T_2E^3(\Delta_{i_3})\|_{\mathbf{S}_2} \ldots \|E^n(\Delta_{i_n})\,T_nE^1(\Delta_{i_1})\|_{\mathbf{S}_2}.$$

Introducing the abbreviation $a_{i_s i_t} = \|E^s(\Delta_{i_s})\,T_sE^t(\Delta_{i_t})\|_{\mathbf{S}_2}$, we obtain

$$\sum_{i_1 i_2 \ldots i_n} a_{i_1 i_2} a_{i_2 i_3} \ldots a_{i_n i_1} \leqslant \left\{ \sum_{i_1 i_2} a_{i_1 i_2}^2 \right\}^{\frac{1}{2}} \cdot \left\{ \sum_{i_2 i_3} a_{i_2 i_3}^2 \right\}^{\frac{1}{2}} \ldots \left\{ \sum_{i_n i_1} a_{i_n i_1}^2 \right\}^{\frac{1}{2}} = \| T_1 \|_{S_2} \| T_2 \|_{S_2} \ldots \| T_n \|_{S_2}.$$

We have proved that $m(\Delta)$ has a weakly bounded S_2 variation. It is well known that to establish denumerable additivity it is sufficient to verify that the following property, called normality, holds: The condition $S_2\text{-}\lim_{s\to\infty} m(\Delta^s) = 0$ holds for any selection of intersecting intervals Δ^s such that $\Delta^s \supset \Delta^{s+1}$. This property obviously follows from the corresponding property of each of the measures E^i.

Let us formulate the above results as a theorem.

THEOREM 1. *Let $T_1, T_2, \ldots, T_n$ be Hilbert – Schmidt operators in a separable Hilbert space H and let $E^1, E^2, \ldots, E_n$ be arbitrary spectral measures in H. Then, the operator-valued function $m(\Delta)$ of the intervals $\Delta = \Delta_1 \times \Delta_2 \times \ldots \times \Delta_n$ defined by*

$$m(\Delta) = E^1(\Delta_1) T_1 E^2(\Delta_2) T_2 \ldots T_{n-1} E^n(\Delta_n)$$

is denumerably additive and has a weakly bounded S_2-variation in Ω.

From intervals, the function $m(\Delta)$ can be extended in the usual manner to the minimal σ-algebra $F \supset F^1 \times F^2 \times \ldots F^n$ of space Ω and generates a denumerably additive measure of weakly bounded S_2-variation. Let us consider the class **M** of functions which are measurable and essentially bounded in this measure, i.e.,

$$(m)\text{-}\sup |\varphi(\lambda)| \equiv \sup_m \operatorname{vrai} |\varphi(\lambda_1, \lambda_2, \ldots, \lambda_n)| < \infty.$$

In view of what has been said above, the functions φ belonging to **M** are integrable with respect to the scalar measure $\mathrm{Sp}\{m(d\lambda) T_n\}$ for any T_n belonging to S_2 and we have the bound

$$\left| \int_\Omega \varphi(\lambda) \mathrm{Sp}\{m(d\lambda) T_n\} \right| \leqslant (m)\text{-}\sup |\varphi(\lambda)| \prod_{k=1}^{n} \| T_k \|_{S_2},$$

which shows that the integral

$$\int_\Omega \varphi(\lambda) \mathrm{Sp}\{m(d\lambda) T_n\}$$

is a linear bounded functional over S_2. It is natural to consider, by definition, that the operator $\Phi(T_1, T_2, \ldots, T_{n-1})$ belonging to S_2 and generated by the above functional is a multiple integral of the function $\varphi(\lambda)$ with respect to measure m in the weak sense, namely,

$$\int \varphi(\lambda) m(d\lambda) = \int \int \ldots \int \varphi(\lambda_1, \lambda_2, \ldots, \lambda_n) dE^1_{\lambda_1} T_1 \ldots dE^n_{\lambda_n} \equiv \Phi(T_1, T_2, \ldots, T_{n-1}).$$

The multiple integral operator defined in this way is a multilinear transformer over the operators $T_1, T_2, \ldots, T_{n-1}$. It obviously possesses all the natural properties of a multiple integral (additivity with respect to the domain and function, uniformity). If $\varphi(\lambda)$ is representable in the form of a product of two functions $\varphi = \varphi_1 \cdot \varphi_2$ depending on two groups of variables $(\lambda_1, \ldots, \lambda_k)$ and $(\lambda_{k+1}, \ldots, \lambda_n)$, then the integral can be represented as

$$\int \varphi(\lambda) m(d\lambda) = \int \varphi_1(\lambda) m_1(d\lambda) \cdot \int \varphi_2(\lambda) m_2(d\lambda);$$

here, we have

$$m_1(\Delta) = E^1(\Delta_1) T_1 \dots E^k(\Delta_k) T_k.$$
$$m_2(\Delta) = E^{k+1}(\Delta_{k+1}) T_{k+1} \dots E^n(\Delta_n).$$

In conclusion, it should be noted that the integral constructed by us is an integral in the Pettis sense (see [6]).

§ 2

In this section we consider the case in which the operators $T_1, T_2, \dots, T_{n-1}$ appearing in expression (2) for $m(\Delta)$ are bounded, the spaces Ω^i are one-dimensional, and energy spectral measure $E^i(\Delta)$ is concentrated on the interval K_{0i} (with $-1 \leq \lambda^i \leq 1$) and is continuous from the left.

Let us take $K_0 = K_{01} \times K_{02} \times \dots \times K_{0n}$. In the case under consideration, the function $m(\Delta)$ does not, in general, have a bounded variation on K_0. Therefore, if we wish to assign a meaning to expression (1), we must adopt a new concept for integration. The approach adopted in this section is essentially that adopted by M. Sh. Birman and M. Z. Solomyak in [1]. According to their definition, the integral of a sufficiently smooth function exists as the limit of integral sums corresponding to a definite class of partitionings called in the following "admissible partitionings."

Admissible Partitionings. The partitioning of the cube K_0 is obtained as the direct product of partitionings along the various coordinate axes. A partitioning Λ^i along coordinate λ^i is generated by a system of points $(\Lambda^i) = \{\lambda^i_k\}$, with $-1 = \lambda^i_1 < \lambda^i_2 < \dots \lambda^i_{k-1} < \lambda^i_k = 1$. The number $r(\Lambda^i) = \sup_s (\lambda^i_{s+1} - \lambda_s)$ will be called the rank of partitioning Λ^i. A partitioning $\tilde{\Lambda}^i$ is said to be an extension of partitioning Λ^i if we have $(\Lambda^i) \subset (\tilde{\Lambda}^i)$; in this case we will write $\Lambda^i \prec \tilde{\Lambda}^i$. The product of two partitionings Λ^i and M^i is the partitioning $N^i = \Lambda^i \circ M^i$, such that $(N^i) = (\Lambda^i) \cup (M^i)$. It is obvious that we have $\Lambda^i \prec \Lambda^i \circ M^i$ for any partitioning M^i. The partitioning $\tilde{\Lambda}^i$ will be called an elementary extension of partitioning Λ^i if

a) $(\Lambda^i) \subset (\tilde{\Lambda}^i)$,

b) every interval $\Delta_s = [\lambda^i_s, \lambda^i_{s+1})$ of partitioning Λ^i contains not more than one point $\lambda^i \subset (\tilde{\Lambda}^i)$.

An admissible partitioning of the cube K_0 is any partitioning Λ obtained as a direct product of partitionings Λ^i, i.e.,

$$\Lambda = \Lambda^1 \times \Lambda^2 \times \dots \times \Lambda^n.$$

The number $r(\Lambda) = \sup_i r(\Lambda^i)$ is called the rank of this partitioning. The partitioning $\tilde{\Lambda} = \tilde{\Lambda}^1 \times \tilde{\Lambda}^2 \times \dots \times \tilde{\Lambda}^n$ will be called the extension of partitioning $\Lambda = \Lambda^1 \times \Lambda^2 \times \dots \times \Lambda^n$ if we have $\Lambda^i \prec \tilde{\Lambda}^i$ for all i; we will denote this by

$$\Lambda \prec \tilde{\Lambda}.$$

Any partitioning of the form

$$\tilde{\Lambda} = \Lambda^1 \times \Lambda^2 \times \dots \times \Lambda^{s-1} \times \tilde{\Lambda}^s \times \Lambda^{s+1} \times \dots \times \Lambda^n,$$

where $\tilde{\Lambda}^s$ is an elementary extension of Λ^s, will be called an elementary extension of Λ. We will write $\tilde{\Lambda} = \Lambda \circ \tilde{\Lambda}^s$.

In the following, an important part will be played by the so-called special sequences of partitionings. Let us define the notion of a special sequence of partitionings along one coordinate λ^s. Let Λ^s and $\tilde{\Lambda}^s$ be any two partitionings and let us assume that $\Lambda^s \prec \tilde{\Lambda}^s$. We can construct a sequence of admissible partitionings

$$\Lambda^s = \Lambda^s_p,\ \Lambda^s_{p+1},\ \ldots,\ \Lambda^s_q,\quad \tilde{\Lambda}^s_q = \Lambda^s_q \circ \tilde{\Lambda}^s,\ \tilde{\Lambda}^s,$$

possessing the following properties:

1. Each successive partitioning except the last is an elementary extension of the proceding one; $\tilde{\Lambda}^s_q$ is an elementary extension of $\tilde{\Lambda}^s$.

2. If an interval Δ is contained in partitionings Λ^s_t, Λ^s_{t+1}, then it is contained in all subsequent partitionings with the possible exception of the last two.

3. The width of every interval Δ of partitioning Λ^s_t, which is subdivided in the transition to partitioning Λ^s_{t+1}, does not exceed 2^{-t+1}.

The method to be used for the construction of such a sequence has been given in [1]. It is important for us that any two partitionings Λ^s, and $\tilde{\Lambda}^s$ (with $\Lambda^s \prec \tilde{\Lambda}^s$) can be connected by a special sequence of partitionings possessing the above three properties. It is obvious that any two partitionings Λ and $\tilde{\Lambda}$ of the cube K_0 (with $\Lambda \prec \tilde{\Lambda}$) can be connected by a sequence of partitionings each of which is an extension of the preceding one along a single coordinate. Further, it is obvious that any two neighboring terms of the above sequence of partitionings can be connected by a chain of special partitionings possessing Properties 1, 2, and 3. A chain of partitionings obtained in this way will be called the special chain of partitionings connecting partitionings Λ and $\tilde{\Lambda}$.

Integral Sums. Let ${}^i\Lambda$ be a partitioning along coordinate λ^i . Let us choose a point $\bar{\lambda}^i_k$ in each segment $\bar{\Delta}^i_k = [\lambda^i_k, \lambda^i_{k+1}]$. The set of all such points will be called the lattice compatible with the partitioning Λ^i, and will be denoted by $\Xi_i(\Lambda^i)$. Let us assume that lattices $\Xi_i(\Lambda^i)$, where i = 1, 2, ..., n, have been constructed. The lattice consisting of the points $\lambda = (\bar{\lambda}^1, \bar{\lambda}^2, \ldots, \bar{\lambda}^n)$, with $\bar{\lambda}^i \in \Xi_i(\Lambda^i)$. will be called the product of lattices Ξ_i (Λ) and will be denoted by

$$\Xi(\Lambda) = \Xi_1(\Lambda^1) \times \Xi_2(\Lambda^2) \times \ldots \times \Xi_n(\Lambda^n).$$

The lattice $\Xi(\Lambda)$ obtained by this construction will be called the regular lattice compatible with partitioning Λ.

Let $\varphi(\lambda_1, \lambda_2, \ldots, \lambda_n) \equiv \varphi(\lambda)$ be a function defined in K_0. The expression

$$S(\Lambda, \Xi(\Lambda), \varphi) = \sum_{\substack{\Delta_{i_1 i_2 \ldots i_n} \in \Lambda \\ \lambda_{i_1 i_2 \ldots i_n} = (\bar{\lambda}^1_{i_1}, \bar{\lambda}^2_{i_2}, \ldots, \bar{\lambda}^n_{i_n}) \in \Xi}} \varphi(\lambda_{i_1 i_2 \ldots i_n}) m(\Delta_{i_1 i_2 \ldots i_n}) \tag{5}$$

will be called the integral sum of this function. If the uniform limit of this expression as the partitioning Λ is made arbitrarily fine exists and this limit is independent of the choice of the lattices $\Xi(\Lambda)$, then the limit will be called the integral of the function $\varphi(\lambda)$ with respect to the set function $m(d\lambda)$, i.e.,

$$(R) \lim_{\substack{(\Lambda, \Xi) \\ r(\Lambda) \to 0}} S(\Lambda, \Xi(\Lambda), \varphi) = \int_{K_0} \varphi(\lambda)\, m(d\lambda). \tag{6}$$

The symbol (Λ, Ξ) underneath the limit sign indicates that a sequence of admissible partitionings and regular lattices has been used in the construction of the integral.

Function Classes. Because the interval function $m(\Delta)$ does not have a bounded variation, the class of functions integrable in the sense of (6) is comparatively narrow. It has been established in [1] that a double integral of the type exists for sufficiently smooth functions. An analogous assertion is valid in the case being considered here. It should be noted that our wish to improve the result forces us to consider weight classes instead of the Sobolev-Slobodetskii spaces used in [2] for the solution of the problem.

Let us first of all consider the case of functions of one variable. Let $\rho(t)$ be a nonnegative function such that

$$\int_0 t^{-3}\rho(t)\,dt < \infty. \tag{7}$$

Making use of V. P. Il'in's formula (see [7]) in the simplest case, we can easily obtain the inequality

$$\sup_{x,\,y\in\Delta} |u(x)-u(y)|^2 \leqslant \iint_{x',\,y'\in\Delta} \frac{|u(x')-u(y')|^2}{\rho(|x'-y'|)}\,dx'\,dy' \int_0^d t^{-3}\rho(t)\,dt, \tag{8}$$

where $d=\sup_{x,\,y\in\Delta}|x-y|$. This inequality is obviously always correct when the right-hand side contains a finite expression. It is also not difficult to see that the class of functions $\{u\}$ defined on $(-1, 1)$ for which the expression

$$\left\{\iint_{-1\leqslant x',\,y'\leqslant 1} \frac{|u(x')-u(y')|^2}{\rho(|x'-y'|)}\,dx'\,dy'\right\}^{\frac{1}{2}} \equiv \|u\|_\rho^0$$

is finite is a linear space with the seminorm $\|u\|_\rho^0$. This class may be converted into a linear normed space $W_{2,\rho}$ if the norm is taken to be

$$\|u\|_\rho = |l(u)| + \|u\|_\rho^0,$$

where $l(u)$ is a linear functional which is not annihilated on the subspace of constants. This norm is equivalent, for example, to the form

$$\|\tilde{u}\|_\rho = \|u\|_{L_2(-1,\,1)} + \|u\|_\rho^0.$$

An immediate consequence of inequality (8) is the theorem of imbedding from $W_{2,\rho}$ into $C(-1, 1)$. The corresponding imbedding operator is completely continuous because all functions $u\in W_{2,\rho}$ have a modulus of continuity $\omega(\delta)$ satisfying the inequality

$$\omega(\delta) \leqslant c\left\{\int_0^\delta t^{-3}\rho(t)\,dt\right\}^{\frac{1}{2}}.$$

In addition to the class $W_{2,\rho}$, let us also consider the class W_2 of functions defined on $(-1, 1)$ and satisfying the condition

$$\int_0 \frac{1}{s}\left\{\sup_{\Delta,\,r(\Delta)\leqslant s}\ \sup_{x_l,\,x_l+h_l\in\bar{\Delta}_l\in\Delta} \sum_l |\Delta_{h_l}u(x_l)|^2\right\}^{\frac{1}{2}} ds < \infty. \tag{9}$$

Here Λ is an arbitrary partitioning of the interval $(-1, 1)$, Δ_i is an interval belonging to partitioning Λ, and $r(\Lambda)$ is the rank of Λ. As will be seen in the following, this function class arises naturally during the investigation of the conditions governing the existence of integral (6). However, the direct verification of condition (9) is very laborious and we will be satisfied to state that $W_{2,\rho} \subset W_2$ if

$$\int_0 t^{-3} \rho(t) \ln^2 t\, dt < \infty. \tag{10}$$

Indeed, in view of inequality (8), a replacement of $\rho(t)$ by $\rho(t)|\ln \alpha t|$, with $\alpha < 1/2$ yields

$$\int_0^a \frac{1}{s} \left\{ \sup_{\Lambda \; r(\Lambda) \leqslant s} \; \sup_{x_l,\, x_l + h_l \in \overline{\Delta}_l \in \Lambda} \sum_l |\Delta_{h_l} u(x_l)|^2 \right\}^{\frac{1}{2}} ds \leqslant$$

$$\leqslant \int_0^a \frac{ds}{s} \left\{ \iint\limits_{\substack{-1 \leqslant x,\, y \leqslant 1 \\ |x-y| \leqslant s}} \frac{|u(x) - u(y)|^2}{\rho(|x-y|)\,|\ln \alpha (x-y)|} dx\, dy \right\}^{\frac{1}{2}} \left\{ \int_0^s t^{-3} \rho(t) |\ln \alpha t|\, dt \right\}^{\frac{1}{2}} \leqslant$$

$$\leqslant \left\{ \int_0^a \frac{ds}{s} \iint\limits_{\substack{-1 \leqslant x,\, y \leqslant 1 \\ |x-y| \leqslant s}} \frac{|u(x) - u(y)|^2}{\rho(|x-y|)\,|\ln \alpha |x-y||} dx\, dy \right\}^{\frac{1}{2}} \left\{ \int_0^a \frac{ds}{s} \int_0^s t^{-3} \rho(t) |\ln \alpha t|\, dt \right\}^{\frac{1}{2}} \leqslant$$

$$\leqslant c \left\{ \iint\limits_{-1 \leqslant x,\, y \leqslant 1} \frac{|u(x) - u(y)|^2}{\rho(|x-y|)} dx\, dy \right\}^{\frac{1}{2}} \cdot \left\{ \int_0^a t^{-3} \rho(t) \ln^2 t\, dt \right\}^{\frac{1}{2}}. \tag{11}$$

In obtaining the last part of this inequality, we have made use of the fact that $\ln \alpha t \sim \ln t$ as $t \to 0$.

Up to now we have considered classes of functions depending on one variable. In the following, we will have to deal with functions of many variables. It will be necessary to consider classes of functions satisfying the condition

$$\|u\|_{W_2(i_1, i_2, \ldots, i_k)} = \int \int \cdots \int \prod_{l \in i_1, \ldots, i_k} \frac{ds^l}{s^l} \sup_{\substack{\lambda^{j_1}, \ldots, \lambda^{j_{n-k}} \\ j \overline{\in} (i_1, \ldots, i_k)}} \times$$

$$\times \left\{ \sup_{\substack{\Pi \Lambda^l,\, r(\Lambda^l) \leqslant s^l \\ l \in i_1, i_1, \ldots, i_k}} \; \sup_{\substack{\lambda^l, \lambda^l + h^l \in \overline{\Delta}^l \\ \Delta^l \in \Lambda^l}} \left[\sum_l \sum_{\Delta^l \in \Lambda^l} |\Delta_{h_{l_1}} \ldots \Delta_{h_{l_k}} \varphi(\lambda_1, \ldots, \lambda_n)|^2 \right] \right\} < \infty \tag{12}$$

instead of condition (9).

As in the case of (9), this condition is also difficult to establish directly. However, we can introduce the class $W_{2,\rho}(i_1, \ldots, i_k) \subset W_2(i_1, \ldots, i_k)$, membership in which can be checked comparatively simply. The class $W_{2,\rho}(i_1, \ldots, i_k)$ consists of the functions satisfying the condition

$$\|u\|^{0}_{\rho,(i_1, \ldots, i_k)} = \sup_{\substack{\lambda_{j_1}, \ldots, \lambda_{j_{n-k}} \\ j \overline{\in} (i_1, \ldots, i_k)}} \left\{ \iint \cdots \iint\limits_{-1 < \lambda_l,\, \lambda_l + h_l < 1} \prod_{l \in i_1, \ldots, i_k} d\lambda^l dh^l \frac{|\Delta_{h_{l_1}} \cdots \Delta_{h_{l_k}} \varphi(\lambda_1, \ldots, \lambda_n)|^2}{\rho(h_{l_1}) \cdots \rho(h_{l_k})} \right\}^{\frac{1}{2}} < \infty. \tag{13}$$

It is easy to prove that a function $u \in W_{2,\rho}(i_1, \ldots, i_k)$ satisfies condition (12) when $\rho(t)$ satisfies (10). In addition, we have the inequality

$$\| u \|_{W_2(i_1, \ldots, i_k)} \leqslant \| u \|^0_{\rho, (i_1, \ldots, i_k)} \left\{ \int_0^2 t^{-3} \rho(t) \ln^2 t \, dt \right\}^k . \tag{14}$$

It is not difficult to see that expression (13) is a seminorm. The class $W_{2,\rho}(i_1, \ldots, i_k)$ can be converted into a normed space if we take expression (13) with some additional terms to be the norm in it.

THEOREM 2. *If function $\varphi(\lambda_1, \lambda_2, \ldots, \lambda_n)$ is bounded and for an arbitrary selection $i_1, i_2, \ldots, i_k \subset (1, 2, \ldots, n)$, which does not simultaneously contain i = 1 and i = n, it satisfies condition (12), then integral (6) exists and represents a bounded operator in H.*

In order to avoid cumbersome expressions, we will give the proof here only for the case n = 3. Let $\varphi(\lambda, \mu, \nu)$ be a bounded function defined in the cube K_0. The partitioning of K_0 along the λ, μ, and ν axes will be denoted by L, M, and N. Any admissible partitioning Λ can be represented as $\Lambda = L \times M \times N$. Every interval Λ is a parallelepiped $\Delta = \Delta_\lambda \times \Delta_\mu \times \Delta_\nu$. The lattice $\Xi(\Lambda)$ is chosen to be regular

$$\Xi(\Lambda) = \Xi_\lambda(L) \times \Xi_\mu(M) \times \Xi_\nu(N).$$

Let $\{\Lambda_i\}$ be an arbitrary sequence of increasingly finer partitionings such that $r(\Lambda_i) \to 0$ as $i \to \infty$. Further, let $S_i \equiv S(\Lambda_i, \Xi(\Lambda_i))$ be any sequence of integral sums over these partitionings. It is necessary to show that the sequence S_i is a fundamental sequence in R. It is sufficient to prove this for the case when the sequence $\{\Lambda_i\}$ is ordered, i.e., when we have $\Lambda_i \prec \Lambda_j$ for i < j. It can be easily seen that the last problem reduces to the evaluation of the difference

$$S(\Lambda, \Xi(\Lambda)) - S(\tilde{\Lambda}, \Xi(\tilde{\Lambda}))$$

under the condition that $\tilde{\Lambda}$ is an extension of Λ along one coordinate only, for example, along λ (the partitionings and lattices along the remaining coordinates are unchanged)

$$\begin{gathered} \Lambda = L \times M \times N, \quad \tilde{\Lambda} = \tilde{L} \times M \times N, \quad L \prec \tilde{L}, \\ \Xi = \Xi_\lambda(L) \times \Xi_\mu(M) \times \Xi_\nu(N), \quad \tilde{\Xi} = \tilde{\Xi}_\lambda(L) \times \Xi_\mu(M) \times \Xi_\nu(N). \end{gathered} \tag{15}$$

Thus, the problem of estimating the difference of integral sums has been reduced to the analogous one-dimensional problem which has been solved in [1]. Theorem 4 of [1] when applied to our case asserts the following: If there exists a nonnegative monotone function $\eta(t)$ satisfying the Dini condition,

$$\int_0 t^{-1} \eta(t) \, dt < \infty,$$

and such that for an arbitrary partitioning L it satisfies the inequality

$$\| S(\Lambda, \Xi(\Lambda)) - S(\Lambda, \Xi'(\Lambda)) \| \leqslant \eta(s) \tag{16}$$

(here $\Lambda = L \times M \times N$, $s = r(L)$, $\Xi = \Xi_\lambda(L) \times \Xi_\mu(M) \times \Xi_\nu(N)$, $\Xi' = \Xi'_\lambda(L) \times \Xi_\mu(M) \times \Xi_\nu(N)$, and $\Xi_\lambda(L)$, $\Xi'_\lambda(L)$ are two arbitrary lattices corresponding to the partitioning), then we have

$$\|S(\Lambda, \Xi) - S(\tilde{\Lambda}, \tilde{\Xi})\| \leqslant 6 \int_0^s t^{-1}\eta(t)\,dt$$

for any pair of partitionings Λ and $\tilde{\Lambda} = \Lambda \circ \tilde{L}$.

In other words, Theorem 4 of [1] asserts that the inequality

$$\|S(\Lambda, \Xi) - S(\tilde{\Lambda}, \tilde{\Xi})\| \leqslant 6 \int_0^s t^{-1}\,dt \sup_{L,\, r(L) \leqslant t} \sup_{\Xi_\lambda(L),\, \Xi'_\lambda(L)} \|S(\Lambda, \Xi) - S(\Lambda, \Xi')\| \tag{17}$$

holds provided that $\tilde{\Lambda} = \Lambda \circ \tilde{L}$, $\Lambda = L \times M \times N$, and $\tilde{L} \succ L$, $r(L) \leqslant s$. Thus, the problem reduces to the evaluation of the expression

$$S(\Lambda, \Xi') - S(\Lambda, \Xi) = \sum_{i,j,k} \Delta_{h_i}\varphi(\lambda_i, \mu_j, \nu_k) F(\Delta_i) X F(\Delta_j) Y G(\Delta_k) \tag{18}$$

for $X, Y \in R'$. Here, we have used the symbol Δ_{h_i} to denote the finite-difference operator for variable λ with step length $h_i = \lambda_i' - \lambda_i$, which is equal to the difference of the first coordinates of corresponding points of Ξ' and Ξ. Let us transform the inner sum

$$\sum_j \Delta_{h_i}\varphi(\lambda_i, \mu_j, \nu_k) F(\Delta_j) \tag{19}$$

in formula (18) into a more useful form.

For this purpose, let us consider the special chain of partitioninings $M_1, \ldots, M_l, \ldots, M_m, M_{m+1} = M_m \circ M, M_{m+2} = M$, connecting the trivial partitioning M_0 of $(-1, 1)$ consisting of the one interval $(-1, 1)$ and the partitioning M. Let us construct the sequence of lattices $\Xi_\mu(M_0) = \Xi'_\mu(M_0)$, $\Xi_\mu(M_1)$, $\Xi'_\mu(M_1)$,

$$\Xi_\mu(M_2), \ldots, \Xi_\mu(M_l), \Xi'_\mu(M_l), \ldots, \Xi_\mu(M_{m+2}),$$

$\Xi'_\mu(M_{m+2}) = \Xi_\mu(M)$ in such a manner that for all s we have

$$\sum_{\substack{\Delta_j \in M_{s-1} \\ \mu'_j \in \Xi'(M_{s-1})}} \Delta_{h_i}\varphi(\lambda_i, \mu'_j, \nu_k) F(\Delta_j) = \sum_{\substack{\Delta_j \in M_s \\ \mu_j \in \Xi(M_s)}} \Delta_{h_i}\varphi(\lambda_i, \mu_j, \nu_k) F(\Delta_j). \tag{20}$$

Condition (20) will be satisfied if we choose $\mu_j = \mu_j'$ in general, except in the intervals M_{k-1} and M_k, that become subdivided, and in fractional intervals $\Delta \in M_{k-1}$ where we take μ_j' to be the point which divides the interval Δ_j into the two intervals Δ_{j1}, $\Delta_{j2} \subset M_k$. The lattice points μ_{j1} and μ_{j2} in the intervals Δ_{j1} and Δ_{j2} will be chosen to coincide with μ_j'. Denoting the finite-difference operator for the variable μ by Δ_{k_j}, we can now rewrite expression (19) as

$$\Delta_{h_i}\varphi(\lambda_i, 0, \nu_k) F(\Delta_0) + \sum_{l=1}^{m+2} \sum_{\Delta_j \in M_l} \Delta_{h_i}\Delta_{k_j}\varphi(\lambda_i, \mu_j, \nu_k) F(\Delta_j). \tag{21}$$

Here, we have $k_j = \mu_j' - \mu_j$, $\mu_j \in \Xi(M_l)$, $\mu_j' \in \Xi'(M_l)$, and $\Delta_0 = (-1, 1)$. Substituting expression (21) into formula (18), we obtain

$$S(\Lambda, \Xi') - S(\Lambda, \Xi) = \sum_{L \times \Delta_0 \times N} \Delta_{h_i}\varphi(\lambda_i, 0, \nu_k) E(\Delta_i) X F(\Delta_0) Y G(\Delta_k) +$$

$$+ \sum_{l=1}^{m+2} \left\{ \sum_{L \times M_l \times N} \Delta_{h_i}\Delta_{k_j}\varphi(\lambda_i, \mu_j, \nu_k) E(\Delta_i) X F(\Delta_j) Y G(\Delta_k) \right\}. \tag{22}$$

Let us now estimate each term of the expression we have obtained

$$\Big|\sum_{L\times\Delta_0\times N}\Delta_{h_l}\varphi(\lambda_l,0,\nu_k)(E(\Delta_l)XF(\Delta_0)YG(\Delta_k)f,g)\Big| =$$
$$= \Big|\sum_L\Big(\sum_N\Delta_{h_l}\varphi(\lambda_l,0,\nu_k)G(\Delta_k)f,\ Y^*F(\Delta_0)X^*E(\Delta_l)g\Big)\Big| \leqslant$$
$$\leqslant\Big\{\sum_L\Big\|\sum_N\Delta_{h_l}\varphi(\lambda_l,0,\nu_k)G(\Delta_k)f\Big\|^2\Big\}^{\frac12}\|Y^*F(\Delta_0)X^*\|\|g\|\leqslant$$
$$\leqslant\Big\{\sum_{L\times N}|\Delta_{h_l}\varphi(\lambda_l,0,\nu_k)|^2\|G(\Delta_k\nu)f\|^2\Big\}^{\frac12}\|Y\|\|X\|\|g\|\leqslant$$
$$\leqslant\sup_{\mu,\nu}\Big\{\sup_{L',\,r(L')\leqslant r(L)}\ \sup_{\lambda_l,\lambda_l+h_l\in\overline{\Delta}_l\in L'}\sum|\Delta_{h_l}\varphi(\lambda_l,\mu,\nu)|^2\Big\}^{\frac12}\|f\|\|g\|\|X\|\|Y\|; \tag{23}$$

$$\Big|\sum_{L\times M\times N}\Delta_{h_l}\Delta_{k_j}\varphi(\lambda_l,\mu_j,\nu_k)(G(\Delta_k)f,\ Y^*F(\Delta_j)X^*E(\Delta_l)g)\Big|\leqslant$$
$$\leqslant\sup_{\nu}\left\{\sup_{\substack{L',\,r(L')\leqslant r(L)\\ M',\,r(M')\leqslant r(M)}}\ \sup_{\substack{\lambda_l,\lambda_l+h_l\in\overline{\Delta}_l\in L'\\ \mu_j,\mu_j+k_j\in\overline{\Delta}_j\in M'}}\sum_{lj}|\Delta_{h_l}\Delta_{k_j}\varphi(\lambda_l,\mu_j,\nu_k)|^2\right\}^{\frac12}\|f\|\|g\|\|X\|\|Y\|. \tag{24}$$

It should be recalled that the special sequence of partitionings has been constructed such that the width of the largest of the intervals of partitioning M_l subject to subdivision is not greater than 2^{-l}. Moreover, we have $k_j = 0$ for all intervals Δ_j belonging to $M_{l-1}\cap M_l$. Let us introduce the abbreviations

$$\sup_{\nu}\left\{\sup_{\substack{L',\,r(L')\leqslant s\\ M',\,r(M')\leqslant t}}\ \sup_{\substack{\lambda_l,\lambda_l+h_l\in\overline{\Delta}_l\in L'\\ \mu_j,\mu_j+k_j\in\overline{\Delta}_j\in M'}}\sum_{lj}|\Delta_{h_l}\Delta_{k_j}\varphi(\lambda_l,\mu_j,\nu)|^2\right\}^{\frac12}\equiv\eta(s,t),$$
$$\sup_{\mu,\nu}\Big\{\sup_{L',\,r(L')\leqslant s}\ \sup_{\lambda_l,\lambda_l+h_l\in\overline{\Delta}_l\in L'}\sum_l|\Delta_{h_l}\varphi(\lambda_l,\mu_j,\nu)|^2\Big\}^{\frac12}\equiv\eta(s).$$

In view of (23), (24), and the above remark, for $l \leq m$ we have

$$\Big\|\sum_{L\times M_l\times N}\Delta_{h_l}\Delta_{k_j}\varphi(\lambda_l,\mu_j,\nu_k)E(\Delta_l)XF(\Delta_j)YG(\Delta_k)\Big\|\leqslant\eta\big(r(L),2^{-l+1}\big). \tag{25}$$

A similar inequality holds when $l = m+1$, provided that we replace $\eta(r(L), 2^{-l+1})$ on the right-hand side by $\eta(r(L), 2^{-m})$. When $l = m+2$, we have

$$\Big\|\sum_{L\times M_{m+2}\times N}\Delta_{h_l}\Delta_{k_j}\varphi(\lambda_l,\mu_j,\nu_k)E(\Delta_l)XF(\Delta_j)YG(\Delta_k)\Big\|\leqslant\eta(r(L),r(M)).$$

Using the inequality

$$\sum_{l=1}^{m}\eta(2^{-l})\leqslant\int_0^1 t^{-1}\eta(t)\,dt,$$

valid for a monotone nonnegative function $\eta(t)$ for all m, we obtain

$$\|S(\Lambda,\Xi)-S(\Lambda,\Xi')\|\leqslant\eta(r(L))+\eta(r(L),r(M))+\int_0^1 s^{-1}\eta(r(L),s)\,ds. \tag{26}$$

If $\varphi(\lambda, \mu, \nu)$ belongs to classes $W(\lambda)$, $W_2(\lambda, \mu)$, then the integral on the right-hand side of (26) converges. Moreover, the integral on the right-hand side of (17) also converges. It follows from this that the sequence of integral sums over partitions belonging to the class under consideration $(\tilde{\Lambda} = \Lambda \circ \tilde{L})$ is a fundamental one. The same argument applies to the case $\tilde{\Lambda} = \Lambda \circ \tilde{M}$. If we have $\tilde{\Lambda} = \Lambda \circ \tilde{N}$, then we must introduce conjugate operators and the problem reduces to the one solved above. It should be noted that the extreme variables λ and ν play special part in our proof: We do not require smoothness with respect to both variables λ and ν simultaneously for integral (6) to exist.

Corollary. If we have $\varphi(\lambda_1, \lambda_2, \ldots, \lambda_n) \subset W_{2,\rho}(i_1, \ldots, i_n)$ for any selection $i_1, i_2, \ldots, i_k \in (1, 2, \ldots, n)$ which does not simultaneously contain i = 1 and i = n, then integral (6) exists. The proof immediately follows from the relationship between the classes $W_2(i_1, i_2, \ldots, i_k)$ and $W_{2,\rho}(i_1, i_2, \ldots, i_k)$. It should be noted that some of the results of [1, 2] follow from our results when we choose $\rho(t)$ in an appropriate manner.

§ 3

In the present section we will investigate the conditions of the existence of the multiple-integral operator

$$\int \varphi(\lambda_1, \lambda_2, \ldots, \lambda_n)\, dE^1_{\lambda_1} T_1\, dE^2_{\lambda_2} T_2 \ldots dE^n_{\lambda_n} \tag{27}$$

in which the T_i are bounded operators, the $E^i_{\lambda_i}$ are multidimensional expansions of unity (or spectral measures), and $\lambda_i = (\lambda^1_i, \lambda^2_i, \ldots, \lambda^{n_i}_i)$. It should be noted that in this case the approach adopted in the last section becomes fruitless. The reason for this is the multidimensionality of the spectral measures E^i. To overcome this difficulty, it is necessary to construct the integral sums with the function replaced on each interval not by a constant, but by a specially constructed function of a simple type. This approach to the definition of integration has been adopted in [2]. It can be described as follows. First of all, we consider the so-called degenerate functions of the form

$$\varphi(\lambda_1, \lambda_2, \ldots, \lambda_n) = \sum_{i_1, i_2, \ldots, i_n}^{N_1, N_2, \ldots, N_n} \varphi_{i_1}(\lambda_1)\varphi_{i_2}(\lambda_2) \ldots \varphi_{i_n}(\lambda_n);$$

here, $\varphi_{i_k}(\lambda_k)$ are bounded functions which are measurable over E^k. It is obvious that an integral of a degenerate function has meaning. If a function $\psi(\lambda_1, \lambda_2, \ldots, \lambda_n)$ is the limit of a sequence of degenerate functions and, in addition, the corresponding sequence of integrals also converges, then the limit of the sequence of integrals is naturally called the integral of $\psi(\lambda_1, \lambda_2, \ldots, \lambda_n)$. Of course, this definition is correct only when this limit is the same for a sufficiently wide class of degenerate-function sequences approximating $\psi(\lambda_1, \lambda_2, \ldots, \lambda_n)$. In this connection, the integral of the degenerate function plays the part of an integral sum.

Let us now proceed to the construction of the sequence of integral sums of integral (27).

Admissible Partitionings. Admissible partitionings along a coordinate λ_i are defined as follows. As the zero-order partitioning Λ_{i0} of the cube K_{i0} we take the regular partitioning whose only interval is the cube K_{i0}. The subsequent partitionings are defined inductively: If partitioning Λ_{ik} has been constructed, then partitioning Λ_{ik+1} is obtained by the subdivision of some of the cubes of Λ_{ik} into 2^{n_i} equal cubes by means of planes drawn parallel to the coordinate planes. The partitioning Λ_{ik+1} constructed in this way is called the elementary extension of partitioning Λ_{ik}. In general, an extension of partitioning Λ_i is the partitioning $\tilde{\Lambda}_i$

obtained from Λ_i by means of a construction of a finite number of successive elementary extensions. Let us now introduce the class of special partitionings. We will say that Λ_{ik+1} is a special elementary extension of partitioning Λ_{ik} if it has been obtained by the subdivision of some cubes of minimum size contained in Λ_{ik} into 2^{n_i} parts. It is obvious that the length of an edge of a minimal cube contained in the partitioning Λ_{ik} of a chain of special partitionings $\Lambda_{i0}, \Lambda_{i1}, \ldots, \Lambda_{ik}$ is equal to 2^{-k+1}. It is not difficult to show that any admissible partitioning can be connected to any of its arbitrary extensions by means of a chain of special partitionings.

Admissible partitionings of the hypercube $K_{10} \times K_{20} \times \ldots \times K_{n0}$ are defined as the products of admissible partitionings of the cubes K_{i0}. The extension of partitioning $\Lambda = \Lambda_1 \times \Lambda_2 \times \ldots \times \Lambda_n$ is the partitioning $\tilde{\Lambda} = \tilde{\Lambda}_1 \times \tilde{\Lambda}_2 \times \ldots \times \tilde{\Lambda}_n$ obtained from Λ by extension along some of the coordinates. As in the previous section, we will use $\Lambda \prec \tilde{\Lambda}$ to denote that $\tilde{\Lambda}$ is an extension of Λ. We define the product of partitionings and elementary extension of a partitioning Λ along the coordinates in the same way as before.

Function Classes and Polynomial Approximation. For the degenerate functions we will take specially constructed piecewise polynomial functions. The problem of the approximation of sufficiently smooth functions by piecewise polynomial functions has been considered in [4]. Let Δ be any cube in R^n. Let us take $u(x) \in W_2^l(\Delta)$, with $2l > n$. We will assume that the $W_2^l(\Delta)$ norm has been defined as follows:

$$\|u\|_{W_2^l(\Delta)} = \sum |l_i(u)| + \|u\|_{L_2^l(\Delta)}.$$

Here, $\{l_i\}$ is a complete system of bounded linear functionals in the subspace L_0^l of functions for which we have $\|u\|_{L_2^l(\Delta)} = 0$. In our case, it is convenient to take the functionals of such a system to be

$$l_{[\varkappa]}u = \int_\Delta x^{(\varkappa)} u(x)\, dx.$$

Here, we have $x^{(\varkappa)} = x_1^{\varkappa_1} x_2^{\varkappa_2} \ldots x_n^{\varkappa_n}$, $\varkappa_i \geqslant 0$, and $\sum \varkappa_i \leqslant [l]$. It is obvious that the polynomials $x^{(\varkappa)}$ form a basis in L_0^l. Let $P_\varkappa^l(x)$ be the corresponding biorthogonal basis in L_0^l [in the sense of the scalar product in $L_2(\Delta)$]. With each function $u(x) \in W_2^l(\Delta)$ let us associate the polynomial

$$P_u^l(x) = \sum_{(\varkappa)} x^{(\varkappa)} \int_\Delta u(\xi) P_\varkappa^{l,\Delta}(\xi)\, d\xi = \int_\Delta P_{(l,\Delta)}(x, \xi)\, u(\xi)\, d\xi \equiv P_{(l,\Delta)} u.$$

The linear operator $P_{(l,\Delta)}$ is called the operator of polynomial approximation in the cube Δ. In view of the inclusion theorem (see [8]), we have

$$\|u - P_{(l,\Delta)} u\|_{C(\Delta)} \leqslant C(l, n)\, d^{\,l - \frac{n}{2}} \|u\|_{L_2^l(\Delta)}. \tag{28}$$

Here, d is the length of an edge of cube Δ.

In the following, we will make use of functions possessing various smoothness properties with respect to distinct variables. Let $\varphi(\lambda_1, \lambda_2, \ldots, \lambda_n)$ be a function defined in the hypercube $\Delta = \Delta_1 \times \Delta_2 \times \ldots \times \Delta_n$ satisfying for $l_i > n_i/2$, where $i = i_1, i_2, \ldots, i_k$, the condition

$$\sup_{\lambda_i,\ i \,\overline{\in}\, i_1, i_2, \ldots, i_k} \|\varphi\|_{L_2^{l_{i_1}, l_{i_2}, \ldots, l_{i_k}}} = \sup_{\lambda_i,\ i \,\overline{\in}\, i_1, i_2, \ldots, i_k} \Bigg\{ \int\int \ldots \int_{\substack{\lambda_i,\, \lambda_i + h_i \in \Delta_i \\ i \in i_1, i_2, \ldots, i_k}} \prod d\lambda_i\, dh_i \times$$

$$\times \frac{\left\| \left[\prod_{l \in i_1, i_2, \ldots, i_k} \left(\Delta_{h_l} D^{[l_l]} \right) \right] \varphi(\lambda_1, \lambda_2, \ldots, \lambda_n) \right\|^2}{\prod_{l \in i_1, i_2, \ldots, i_k} h_l^{n_l + 2(l_l - [l_l])}} \Bigg\}^{\frac{1}{2}} < \infty. \tag{29}$$

It is not difficult to see that the expression $\|\varphi\|_{L_2}^{l_{i_1}, l_{i_2}, \ldots, l_{i_k}}$ is a seminorm. The class of functions satisfying (29) can be converted into a normed space if we take (29) supplemented by some other terms to be the norm. It is not difficult to show that the functions satisfying condition (29) can be approximated by polynomials in the following sense:

$$\sup_{\lambda_l, l = i_1, i_2, \ldots, i_k} \left| \left\{ \prod_{l \in i_1, i_2, \ldots, i_k} [1 - P_{(l_l, \Delta_l)}] \right\} \varphi \right| \leqslant C \prod_{l = i_1, i_2, \ldots, i_k} d_l^{l_l - \frac{n_l}{2}} \|\varphi\|_{L_2}^{l_{i_1}, l_{i_2}, \ldots, l_{i_k}}. \tag{30}$$

<u>Degenerate Functions and Integral Sums.</u> Let $\Lambda = \Lambda_1 \times \Lambda_2 \times \ldots \times \Lambda_n$ be an admissible partitioning of $\Delta = \Delta_1 \times \Delta_2 \times \ldots \times \Delta_n \in \Lambda$. We define the degenerate function φ_Λ corresponding to the given partition by

$$\varphi_\Lambda(\lambda) = \prod_{l=1}^{n-1} P_{(l_l, \Delta_l)} \varphi, \quad \lambda \in \Delta.$$

It is obvious that $\varphi_\Lambda(\lambda)$ is a piecewise polynomial function in the variables $\lambda_1, \lambda_2, \ldots, \lambda_{n-1}$. Moreover, the condition

$$\varphi_\Lambda(\lambda) - (\varphi_\Lambda)_{\tilde{\Lambda}}(\lambda) = 0 \tag{31}$$

holds for any $\tilde{\Lambda} > \Lambda$.

The integral of the degenerate function φ_Λ with respect to $m(d\lambda)$ will be called the integral sum over the partitioning

$$S_\Lambda = \int \varphi_\Lambda(\lambda)\, m(d\lambda).$$

THEOREM 3. Let $\varphi(\lambda)$ be a function which is defined in the hypercube $K_0 = K_{01} \times K_{02} \times \ldots \times K_{0n}$ and which satisfies condition (29) in K_0 for any selection $i_1, i_2, \ldots, i_n$, with $i \neq n$ (including an empty selection). Then, as the partitionings are made arbitrarily fine, the sequence of integral sums converges in R to a limit which is independent of the choice of the sequence of admissible partitionings being made increasingly fine.

The above limit will be called the integral of $\varphi(\lambda)$ with respect to $m(d\lambda)$ and will be denoted by

$$\int \varphi(\lambda)\, m(d\lambda).$$

We will prove Theorem 3 here for the case of three variables. This will enable us to avoid a cumbersome notation. The general case can be proved in an analogous manner.

Let $\lambda = (\lambda^1, \lambda^2, \ldots, \lambda^{n_\lambda})$, $\mu = (\mu^1, \mu^2, \ldots, \mu^{n_\mu})$, and $\nu = (\nu^1, \nu^2, \ldots, \nu^{n_\nu})$ be variables whose ranges lie in the cubes $K_{0\lambda}$, $K_{0\mu}$, and $K_{0\nu}$, respectively. Let us consider the function $\varphi(\lambda, \mu, \nu)$ which satisfies conditions (29) for $l_\lambda > n_\lambda/2$ and $l_\mu > n_\mu/2$,

$$\sup_{\mu,\nu} \iint\limits_{\lambda,\,\lambda+h \in K_{0\lambda}} \frac{|\Delta_h D^{[l_\lambda]}\varphi(\lambda,\mu,\nu)|^2}{h^{n_\lambda+2(l_\lambda-[l_\lambda])}}\, d\lambda\, dh < \infty,$$

$$\sup_{\lambda,\nu} \iint\limits_{\mu,\,\mu+k \in K_{0\mu}} \frac{|\Delta_k D^{[l_\mu]}\varphi(\lambda,\mu,\nu)|^2}{k^{n_\mu+2(l_\mu-[l_\mu])}}\, d\mu\, dh < \infty, \tag{32}$$

$$\sup_{\nu} \iint\limits_{\mu,\,\mu+k \in K_{0\mu}} d\mu\, dk \iint\limits_{\lambda,\,\lambda+h \in K_{0\lambda}} \frac{|\Delta_k \Delta_h D^{[l_\mu]} D^{[l_\lambda]}\varphi(\lambda,\mu,\nu)|^2}{h^{n_\lambda+2(l_\lambda-[l_\lambda])} k^{n_\mu+2(l_\mu-[l_\mu])}}\, d\lambda\, dh < \infty.$$

As in the preceding section, we must show that the sequence of integral sums over any sequence of increasingly finer admissible partitionings is fundamental in R. It is not difficult to see that the verification of the conditions of convergence of a sequence of integral sums reduces to the evaluation of the inequality

$$\|S_{\tilde{\Lambda}} - S_\Lambda\| \leqslant \eta(r),\ r = r(\Lambda),\ \tilde{\Lambda} > \Lambda, \tag{33}$$

where the function $\eta(r)$ satisfies the Dini condition. The proof of this assertion is obtained analogously to the proof of Theorem 5 of [1] with some simplifications.

In the following, our calculations are devoted to the derivation of inequality (33). It is obvious that it will be sufficient to consider the case when $\tilde{\Lambda}$ is an extension of Λ along one coordinate only, say, λ. Let L, M, N be partitionings along the coordinates λ, μ, and ν, respectively, and let Λ = L × M × N, $\tilde{\Lambda} = \Lambda \circ \tilde{L}$, and $\tilde{L} > L$. Further, let the interval function $m(\Delta)$ be of the form

$$m(\Delta) = E(\Delta_\lambda) X F(\Delta_\mu) Y G(\Delta_\nu),$$

where $\Delta = \Delta_\lambda \times \Delta_\mu \times \Delta_\nu$, $X, Y \in R$, and E, F, and G are spectral measures. Let us consider the difference

$$S_{\tilde{\Lambda}} - S_\Lambda = \int\limits_{K_0} (\varphi_{\tilde{\Lambda}} - \varphi_\Lambda)\, dE_\lambda X\, dF_\mu Y\, dG_\nu. \tag{34}$$

Let us transform the integral in formula (34) into a more convenient form. Let P_L, P_M denote the operators of piecewise polynomial approximation on the intervals of partitionings L and M, respectively, so that we have

$$\varphi_{\tilde{\Lambda}} - \varphi_\Lambda = (P_{\tilde{L}} P_M - P_L P_M)\varphi = (P_{\tilde{L}} - P_L) P_M \varphi.$$

Now, let $M_0, M_1, \ldots, M_m = M$ be a special chain of partitionings connecting M to the trivial partitioning M_0. By definition, the length of the edge of the minimum interval of partitioning M_t is 2^{-t+1}. Further, let $L_k = L, L_{k+1}, \ldots L_{\tilde{k}} = \tilde{L}$ be a special chain of partitionings connecting partitionings L and $\tilde{L}$. We will assume that the length of the edges of intervals $\Delta_\lambda \in L_s$, subject to subdivision in the transition to partitioning L_{s+1} is equal to 2^{-s+1}. This enumeration of L_s is possible because the sequence $\{L_s\}$ is a special sequence of partitionings. Making use of the formulas

$$P_{\tilde{L}} - P_L = \sum_{s=k}^{\tilde{k}-1} (P_{L_{s+1}} - P_{L_s}),$$

$$P_M - P_{M_0} = \sum_{t=0}^{m-1} (P_{M_{t+1}} - P_{M_t}),$$

we obtain

$$\varphi_{\tilde{\Lambda}} - \varphi_{\Lambda} = \sum_{s=k}^{\tilde{k}-1} (P_{L_{s+1}} - P_{L_s}) P_{M_0}\varphi + \sum_{s=k}^{\tilde{k}-1} \sum_{t=0}^{m-1} (P_{L_{s+1}} - P_{L_s})(P_{M_{t+1}} - P_{M_t})\varphi. \tag{35}$$

It should be noted that the operator $P_{L_{s+1}} - P_{L_s}$ is in fact the operator of annihilation on all intervals $\Delta_\lambda \in L_s \cap L_{s+1}$. The edge of each interval $\Delta_\lambda \in L_{s+1}$ on which we have $P_{L_{s+1}} - P_{L_s} \neq 0$ is of length 2^{-s}. An analogous assertion holds for the intervals of partitioning M_t. In the following, we will need an estimate of every term of (35). Let us take $\lambda \in \Delta'_\lambda \in L_{s+1}$, $\Delta'_\lambda \subset \Delta_\lambda \in L_s$, $\Delta'_\lambda \neq \Delta_\lambda$, and $\mu \in M_0$. In view of (28), we then have

$$\|(P_{L_{s+1}} - P_{L_s}) P_{M_0}\varphi\| \leqslant \|(P_{L_{s+1}} - 1) P_{M_0}\varphi\| + \|(P_{L_s} - 1) P_{M_0}\varphi\| \leqslant$$

$$\leqslant C(2^{-s})^{l_\lambda - \frac{n_\lambda}{2}} \|P_{M_0}\varphi\|_{L_2^{l_\lambda}(\Delta_\lambda)} + C(2^{-s+1})^{l_\lambda - \frac{n_\lambda}{2}} \|P_{M_0}\varphi\|_{L_2^{l_\lambda}(\Delta_\lambda)} \leqslant C(2^{-s})^{l_\lambda - \frac{n_\lambda}{2}} \|P_{M_0}\varphi\|_{L_2^{l_\lambda}(\Delta_\lambda)}. \tag{36}$$

Analogously, in the case of the terms of the summation in (35) with

$$\lambda \in \Delta'_\lambda \in L_{s+1},\ \Delta'_\lambda \subset \Delta\lambda \in L_s,\ \Delta'_\lambda \neq \Delta_\lambda,$$
$$\mu \in \Delta'_\mu \in M_{t+1},\ \Delta'_\mu \subset \Delta_\mu \in M_t,\ \Delta'_\mu \neq \Delta\mu,$$

the following estimate holds because of (30):

$$\|(P_{L_{s+1}} - P_{L_s})(P_{M_{t+1}} - P_{M_t})\varphi\| \leqslant C(2^{-s})^{l_\lambda - \frac{n_\lambda}{2}} (2^{-t})^{l_\mu - \frac{n_\mu}{2}} \|\varphi\|_{L_{2(\Delta_\lambda \times \Delta_\mu)}^{l_\lambda, l_\mu}}. \tag{37}$$

It is obvious that every term of the summation being considered can be represented on each interval of partitioning Λ as

$$\sum_{\alpha, \beta \leqslant [l_\lambda], [l_\mu]} (\lambda - \bar{\lambda})^\alpha (\mu - \bar{\mu})^\beta a_{\alpha\beta}(\bar{\lambda}, \bar{\mu}, \nu) \equiv P_\nu(\mu, \lambda), \tag{38}$$

where $\bar{\lambda}$ and $\bar{\mu}$ are the central points of the intervals Δ_λ and Δ_μ, respectively. It is also not difficult to see that for every polynomial of the form of (38) we have

$$|a_{\alpha\beta}(\nu)|\, d_\lambda^\alpha d_\mu^\beta \leqslant C \sup_{\lambda, \mu \in \Delta_\lambda \times \Delta_\mu} |P_\nu(\lambda, \mu)| \tag{39}$$

with C a constant depending only on l_λ, l_μ, n_λ, and n_μ.

Let us now consider the bilinear form of the operator (34) and substitute the sum (35) for the integrand. It should be noted that the integrations over the intervals Δ_ν can be combined, so that we can integrate immediately over $K_{0\nu}$.

Performing these transformations, we obtain

$$(S_{\tilde{\Lambda}}f, g) - (S_\Lambda f, g) = \sum_{s=k}^{\tilde{k}-1} \Bigg\{ \sum_{\substack{\Delta_\lambda \in L_{s+1} \\ \Delta_\mu = K_{0\mu}}} \int_{K_{0\nu}} \int_{\Delta_\lambda} \int_{\Delta_\mu} (P_{L_{s+1}} - P_{L_s}) P_{M_0}\varphi(\lambda, \mu, \nu) \times$$

$$\times (dE_\lambda X\, dF_\mu Y\, dG_\nu f,\, g)\Bigg\} + \sum_{s=k}^{k-1}\sum_{t=0}^{m-1}\Bigg\{\sum_{\substack{\Delta_\lambda \in L_{s+1}\\ \Delta_\mu \in M_{t+1}}} \int_{\Delta_\lambda}\int_{\Delta_\mu}\int_{K_{0\nu}} (P_{L_{s+1}} - P_{L_s})\times$$

$$\times (P_{M_{t+1}} - P_{M_t})\,\varphi(\lambda, \mu, \nu)\,(dE_\lambda X\, dF_\mu Y\, dG_\nu f,\, g)\Bigg\}. \tag{40}$$

Let us now estimate the terms in the braces in the above expression. For example, let us consider the term in the braces in the second sum, namely,

$$\sum_{\substack{\Delta_\lambda \in L_{s+1}\\ \Delta_\mu \in W_{t+1}}} \int_{\Delta_\lambda}\int_{\Delta_\mu}\int_{K_{0\nu}} (P_{L_{s+1}} - P_{L_s})(P_{M_{t+1}} - P_{M_t})\;\varphi(\lambda, \mu, \nu)\,(dE_\lambda X\, dF_\mu Y\, dG_\nu f,\, g).$$

Representing the integrand in the form of (38), we find that this term becomes

$$\sum_{\substack{\Delta_\lambda \in L_{s+1}\\ \Delta_\mu \in M_{t+1}}} \sum_{\alpha\beta}\left(\int_{\Delta_\lambda}(\lambda-\bar\lambda)^\alpha\, dE_\lambda X \int_{\Delta_\mu}(\mu-\bar\mu)^\beta\, dF_\mu Y \int_{K_{0\nu}} a_{\alpha\beta}(\bar\lambda, \bar\mu, \nu)\, dG_\nu f,\, g\right). \tag{41}$$

The outer summation here extends, in fact, only over the intervals $\Delta_\lambda \in L_{s+1}$ and $\Delta_\mu \in M_{t+1}$ that are not contained in L_s and M_t, respectively; $\bar\lambda$ and $\bar\mu$ are the central points of Δ_λ and Δ_μ. Let us estimate expression (41).

$$\left|\sum_{\substack{\Delta_\lambda \in L_{s+1}\\ \Delta_\mu \in M_{t+1}}} \sum_{\alpha\beta}\left(\int_{K_{0\nu}} a_{\alpha\beta}(\bar\lambda, \bar\mu, \nu)\, dG_\nu f,\, Y^*\int_{\Delta_\mu}(\mu-\bar\mu)^\beta\, dF_\mu X^* \int_{\Delta_\lambda}(\lambda-\lambda)^\alpha\, dE_\lambda g\right)\right| \leqslant \sum_{\alpha\beta}\sum_{\Delta_\lambda \in L_{s+1}} \times$$

$$\times\left(\sum_{\Delta_\mu \in M_{t+1}}\left\|\int_{K_{0\nu}} a_{\alpha\beta}(\bar\lambda, \bar\mu, \nu)\, dG_\nu f\right\|^2\right)^{\frac12}\left(\sum_{\Delta_\mu \in M_{t+1}} \|Y^*\|^2 d^{2\beta}(\Delta_\mu)\left\|F(\Delta_\mu)\int_{\Delta_\lambda}(\lambda-\bar\lambda)^\alpha\, dE_\lambda g\right\|^2\right)^{\frac12} =$$

$$= \sum_{\alpha\beta}\sum_{\Delta_\lambda \in L_{s+1}}\left(\sum_{\Delta_\mu \in M_{t+1}}\int_{K_{0\nu}} |a_{\alpha\beta}(\bar\lambda, \bar\mu, \nu)|^2 (dG_\nu f, f)\right)^{\frac12} \|Y\|\, d^\beta(\Delta_\mu)\left\|\int_{\Delta_\lambda}(\lambda-\bar\lambda)^\alpha\, dE_\lambda g\right\| \leqslant$$

$$\leqslant \sum_{\alpha\beta}\left\{\sum_{\substack{\Delta_\lambda \in L_{s+1}\\ \Delta_\mu \in M_{t+1}}}\int_{K_{0\nu}} |a_{\alpha\beta}(\bar\lambda, \bar\mu, \nu)|^2\, d(G_\nu f, f)\right\}^{\frac12} d^\alpha(\Delta_\lambda)\, d^\beta(\Delta_\mu)\, \|g\|.$$

Taking (37) and (38) into account, we obtain

$$\sum_{\substack{\Delta_\lambda \in L_{s+1}\\ \Delta_\mu \in M_{t+1}}} |a_{\alpha\beta}(\bar\lambda, \bar\mu, \nu)|^2 d^{2\alpha}(\Delta_\lambda)\, d^{2\beta}(\Delta_\mu) \leqslant C\,(2^{-s})^{2l_\lambda - n_\lambda}(2^{-t})^{2l_\mu - n_\mu}\|\varphi\|^2_{L_2^{l_\lambda l_\mu}(K_{0\lambda}\times K_{0\mu})}.$$

Here, C is a constant that depends on l_λ, l_μ, n_λ, and n_μ.

Making use of the last inequality, we can easily see that the estimate of the absolute magnitude of expression (41) is given by

$$C\cdot 2^{-s\left(l_\lambda - \frac{n_\lambda}{2}\right)}2^{-t\left(l_\mu - \frac{n_\mu}{2}\right)}\sup_\nu \|\varphi\|_{L_2^{l_\lambda l_\mu}(K_{0\lambda}\times K_{0\mu})}\|f\|\,\|g\|.$$

Here, as above, C is a constant that depends only on l_λ, l_μ, n_λ, and n_μ.

An analogous estimate can be obtained for the terms of the first sum in expression (40), namely,

$$\left|\sum_{\substack{\Delta_\lambda \in L_{s+1}\\ \Delta_\mu = K_{0\mu}}} \int_{\Delta_\lambda}\int_{\Delta_\mu}\int_{K_{0\nu}} (P_{L_{s+1}} - P_{L_s}) P_{M_0}\varphi(\lambda,\mu,\nu)(dE_\lambda X\, dF_\mu Y\, dG_\nu f, g)\right| \leqslant C 2^{-s\left(l_\lambda - \frac{n_\lambda}{2}\right)} \sup_{\nu,\mu} \|\varphi\|_{L_2^{l_\lambda}(K_{0\lambda})} \|f\|\|g\|.$$

The inequality for the norm of $S_{\tilde\Lambda} - S_\Lambda$ that we require now follows directly from the above estimates and we have

$$\|S_{\tilde\Lambda} - S_\Lambda\| \leqslant C\left\{\sum_{s=k}^{\tilde k-1} 2^{-s\left(l_\lambda - \frac{n_\lambda}{2}\right)} + \sum_{t=0}^{m}\sum_{s=k}^{\tilde k-1} 2^{-s\left(l_\lambda - \frac{n_\lambda}{2}\right)} 2^{-t\left(l_\mu - \frac{n_\mu}{2}\right)}\right\}\left\{\sup_{\nu,\mu}\|\varphi\|_{L_2^{l_\lambda}(K_{0\lambda})} + \sup_{\nu}\|\varphi\|_{L_2^{l_\lambda l_\mu}(K_{0\lambda}\times K_{0\mu})}\right\} \leqslant$$

$$\leqslant C\cdot 2^{-k\left(l_\lambda - \frac{n_\lambda}{2}\right)}\left\{\sup_{\nu,\mu}\|\varphi\|_{L_2^{l_\lambda}(K_{0\lambda})} + \sup_{\nu}\|\varphi\|_{L_2^{l_\lambda l_\mu}(K_{0\lambda}\times K_{0\mu})}\right\} \leqslant$$

$$\leqslant C[r(\Lambda)]^{l_\lambda - \frac{n_\lambda}{2}}\left\{\sup_{\nu,\mu}\|\varphi\|_{L_2^{l_\lambda}(K_{0\lambda})} + \sup_{\nu}\|\varphi\|_{L_2^{l_\lambda l_\mu}(K_{0\lambda}\times K_{0\mu})}\right\}. \qquad (42)$$

It can be seen from inequality (42) that we can take the function $\eta(r)$ in estimate (33) to be cr^{ε}, with $\varepsilon = \max\left(l_\lambda - \frac{n_\lambda}{2}\right),\ \left(l_\mu - \frac{n_\mu}{2}\right)$. This function obviously satisfies Dini's condition. The theorem has been proved.

The author would like to express his deep gratitude to M. Sh. Birman and M. Z. Solomyak for their attention to the work.

Literature Cited

1. M. Sh. Birman and M. Z. Solomyak, "Stieltjes double-integral operators," in: Topics in Mathematical Physics, Vol. 1, M. Sh. Birman (editor), Consultantus Bureau, New York (1967), p. 25.
2. M. Sh. Birman and M. Z. Solomyak, "Stieltjes double-integral operators. II," in: Topics in Mathematical Physics, Vol. 2, M. Sh. Birman (editor), Consultants Bureau, New York (1968), p. 19.
3. M. Sh. Birman and M. Z. Solomyak, "On estimates of the singular numbers of integral operators. II," Vestnik LGU, No. 13 (1967).
4. M. Sh. Birman and M. Z. Solomyak, "Piecewise polynomial approximations of functions belonging to classes W_p^α," Matem. Sbornik, Vol. 73, No. 3 (115), pp. 50-74 (1967).
5. M. Z. Solomyak and V. V. Sten'kin, "On one class of Stieltjes multiple-integral operators," Present volume, p. 99.
6. E. Hille and R. Phillips, Functional Analysis and Semigroups [Russian translation], IL, Moscow (1951), pp. 60-64.
7. V. P. Il'in and V. A. Solonnikov, "On some properties of differentiable functions of many variables," Dokl. Akad. Nauk SSSR, Vol. 136, No. 3, pp. 538-544 (1961).
8. S. M. Nikol'skii, "On theorems of imbedding, extension, and approximation," Uspekhi Matem. Nauk, Vol. 16, No. 3 (101), pp. 63-114 (1961).

ON ONE CLASS OF STIELTJES MULTIPLE-INTEGRAL OPERATORS

M. Z. Solomyak and V. V. Sten'kin

§ 1. Introduction

The concept of a Stieltjes multiple-integral operator was first introduced by Yu. L. Daletskii and S. G. Krein [1] in connection with certain aspects of analytical perturbation theory. In a number of articles [2-5], M. Sh. Birman and M. Z. Solomyak developed the theory of double-integral operators which they found to be closely related to a variety of problems in analysis.

Without introducing rigorous definitions at this stage, let us give a formal description of multiple-integral operators. Let $E_\lambda^{(1)}, \ldots, E_\lambda^{(n)}$ be orthogonal expansions of unity in a Hilbert space **H**, let $T_1, \ldots, T_{n-1}$ be bounded operators in **H**, and let $\varphi(\lambda_1, \ldots, \lambda_n)$ be a scalar complex function. An integral of the form*

$$Q = \underbrace{\int \cdots \int}_{n} \varphi(\lambda_1, \ldots, \lambda_n)\, dE_{\lambda_1}^{(1)} T_1\, dE_{\lambda_2}^{(2)} \ldots T_{n-1}\, dE_{\lambda_n}^{(n)} \tag{1.1}$$

is called a Stieltjes n-tuple-integral operator.

The fundamental problem of the theory of integral operators consists in the clarification of the meaning of integral (1.1) and the determination of the properties of the operator Q defined by it. The answer to these questions naturally depends on the class of operators $T_1, \ldots, T_{n-1}$ and the class of functions φ considered in the problem. The first results obtained in this direction are contained in [1]. The case n = 2 has been investigated in detail in [2-5]. A number of important results has been obtained by B. S. Pavlov [6] for the case of arbitrary n.

In addition to results of a general nature, it is of interest to establish facts concerning integrals of one of the special types. Thus, a special part is played in perturbation theory by integrals of the form of (1.1) in which the function φ is a difference relation of order $n-1$ for a function of one variable. Integrals of functions of this type have been considered in [1] where

*We do not consider here the more general case in which integration is performed with respect to arbitrary (not necessarily one-dimensional) spectral measures.

existence conditions were obtained for them (with a suitable definition of the integral).* Double-integral operators were studied in [2-3] for the case in which the function φ has the form

$$\varphi(\lambda_1, \lambda_2) = \frac{f(\lambda_2) - f(\lambda_1)}{\lambda_2 - \lambda_1}. \tag{1.2}$$

It was possible to obtain for integrals of a function of the form of (1.2) a stronger result than in the case of an arbitrary function φ. At the same time, the corresponding result of [1] was also strengthened.

The principal aim of the present article is the generalization of the result obtained in [2, 3] for double integrals of functions of the form of (1.2) to multiple integrals of functions which represent difference relations of higher orders. The proof of the corresponding result is contained in Section 3.

Our constructions are based on a definition of integral (1.1) suitable for a special class of functions. This definition and associated results are discussed in Section 2. It should be noted that the function class considered by us intersects the class for which n-tuple integrals have been defined by B. S. Pavlov [6], but is not covered by it. It can be easily shown that for functions belonging to the intersection of these classes the two definitions are consistent.

§ 2. On the Definition of the Integral

Let us discuss the assumptions which will be used in this article for the investigation of integrals of the form of (1.1). Let each of the expansions of unity $E_\lambda^{(j)}$ with $j = 1,\dots, n$, be constant outside an interval $[a, b]$ which is common to all numbers j. Let A_j, with $j = 1,\dots, n$, denote the self-adjoint operator which generates the $E_\lambda^{(j)}$.

If integral (1.1) exists in some sense, then it associates an operator Q with function φ and operators $T_1,\dots, T_{n-1}$. Let us agree to write

$$Q = J(\varphi;\ T_1, \dots, T_{n-1});$$

if the operators $T_1,\dots, T_{n-1}$ remain fixed throughout some discussion, then we will use the shorter form

$$Q = J\varphi.$$

Let us first of all consider functions φ of the form

$$\varphi(\lambda_1, \dots, \lambda_n) = g_1(\lambda_1)\dots g_n(\lambda_n) \quad (\lambda_j \in [a, b];\ j = 1, \dots, n), \tag{2.1}$$

where each function g_j, with $j = 1,\dots, n$, is Borel measurable and bounded. In this case, integral (1.1) is naturally to be understood as the operator

$$Q = \int g_1(\lambda_1)\,dE_{\lambda_1}^{(1)} \cdot T_1 \cdot \int g_2(\lambda_2)\,dE_{\lambda_2}^{(2)} \cdot \dots \cdot T_{n-1} \cdot \int g_n(\lambda_n)\,dE_{\lambda_n}^{(n)},$$

i.e.,

$$J(\varphi;\ T_1, \dots, T_{n-1}) = g_1(A_1)\,T_1 g_2(A_2)\dots T_{n-1} g_n(A_n). \tag{2.2}$$

Relation (2.2) yields the inequality

$$\| J(\varphi;\ T_1, \dots, T_{n-1})\| \leqslant \sup_{\lambda \in [a,b]} |g_1(\lambda)| \dots \sup_{\lambda \in [a,b]} |g_n(\lambda)| \cdot \|T_1\| \dots \|T_{n-1}\|. \tag{2.3}$$

---

* An exact formulation of this result is given in Section 3, below.

Let us assume that the functions $g_1, \ldots, g_n$ are continuous on the interval $[a, b]$. In this case [still only for functions of the form of (2.1)], the integral (1.1) can also be regarded as a limit of integral sums.

Let Ξ be a partitioning of the segment $[a, b]$ into nonoverlapping intervals $\Delta_1, \ldots, \Delta_\nu$ and let $h(\Xi)$ be the largest length of these intervals. Let us consider the integral sums

$$S \equiv S(\varphi;\ T_1, \ldots, T_{n-1};\ \Xi;\ \xi_1, \ldots, \xi_\nu) = \sum_{l_1, \ldots, l_n = 1}^{\nu} \varphi(\xi_{l_1}, \ldots, \xi_{l_n}) E^{(1)}(\Delta_{l_1}) T_1 E^{(2)}(\Delta_{l_2}) \ldots T_{n-1} E^{(n)}(\Delta_{l_n}) \tag{2.4}$$

$$(\xi_l \in \Delta_l;\quad l = 1, \ldots, \nu).$$

If φ is a function of the form of (2.1), then, as can be easily seen, the limit of sums of the form of (2.4)* exists as $h(\Xi) \to 0$ in the operator norm and is equal to $Q = J(\varphi;\ T_1, \ldots, T_{n-1})$, i.e.,

$$\lim_{h(\Xi) \to 0} \| S(\varphi;\ T_1, \ldots, T_{n-1};\ \Xi;\ \xi_1, \ldots, \xi_\nu) - J(\varphi;\ T_1, \ldots, T_{n-1}) \| = 0. \tag{2.5}$$

Because of linearity, definition (2.2) of the integral can be extended to finite linear combinations of functions of the form of (2.1). It is obvious that the limit (2.5) is preserved. It follows from this that the meaning of the integral depends on the function φ itself and not on its representation as a linear combination.

Let us now assume that φ can be expanded in a series

$$\varphi(\lambda_1, \ldots, \lambda_n) = \sum_{m=1}^{\infty} g_{m,1}(\lambda_1) \ldots g_{m,n}(\lambda_n) \tag{2.6}$$

such that we have

$$\sum_{m=1}^{\infty} \sup_{\lambda \in [a, b]} |g_{m,1}(\lambda)| \cdot \ldots \cdot \sup_{\lambda \in [a, b]} |g_{m,n}(\lambda)| < \infty. \tag{2.7}$$

Let us take

$$\varphi_m(\lambda_1, \ldots, \lambda_n) = g_{m,1}(\lambda_1) \ldots g_{m,n}(\lambda_n) \quad (m = 1, 2, \ldots).$$

Because of inequality (2.3) and conditions (2.7), the series

$$\sum_{m=1}^{\infty} J(\varphi_m;\ T_1, \ldots, T_{n-1})$$

converges in the norm to an operator. The limiting operator is then naturally called the n-tuple-integral operator constructed from the function φ of the form of (2.6) and the operators $T_1, \ldots, T_{n-1}$. However, we have still to establish that this limit is independent of the representation of the function φ as a series.

We will prove this assertion for a somewhat narrower class of functions φ on the basis of the following lemma.†

*We can also consider integral sums of a more general type, although this would lead to an appreciable complication of the notation. On the other hand, for the purposes of this article it is sufficient to consider sums of the special form of (2.4).

† The procedure involving a lemma of the type of Lemma 1 for proving uniqueness has been adopted from K. Töllner's article [7].

LEMMA 1. Let the function φ be representable in the form of the series (2.6), the functions $g_{m,j}$ with $j = 1, \ldots, n$ and $m = 1, 2, \ldots$ being continuous, and let condition (2.7) be satisfied. Then, the integral sums (2.4) converge in the norm to the operator

$$Q = \sum_{m=1}^{\infty} J(\varphi_m; T_1, \ldots, T_{n-1}).$$

Proof. Let us introduce the following abbreviations:

$$M_{m,j} = \max_{\lambda \in [a,b]} |g_{m,j}(\lambda)| \quad (j = 1, \ldots, n;\ m = 1, 2, \ldots);$$

$$M = \sum_{m=1}^{\infty} M_{m,1} \ldots M_{m,n}.$$

We will show that under the conditions of the lemma, the integral sums corresponding to any partitioning Ξ of the interval $[a, b]$ and an arbitrary selection of points $\xi_1, \ldots, \xi_\nu$ satisfy the inequality

$$\| S(\varphi; T_1, \ldots, T_{n-1}; \Xi; \xi_1, \ldots, \xi_\nu) \| \leqslant M \| T_1 \| \ldots \| T_{n-1} \|. \tag{2.8}$$

Indeed, let us introduce the step-functions

$$\tilde{g}_{m,j}(\lambda) = g_{m,j}(\xi_l) \quad (\lambda \in \Delta_l;\ l = 1, \ldots, \nu)$$
$$(j = 1, \ldots, n;\ m = 1, 2, \ldots)$$

and let us transform the sum (2.4)

$$S = \sum_{l_1, \ldots, l_n = 1}^{\nu} \sum_{m=1}^{\infty} g_{m,1}(\xi_{l_1}) \ldots g_{m,n}(\xi_{l_n})\ E^{(1)}(\Delta_{l_1}) T_1 E^{(2)}(\Delta_{l_2}) \ldots T_{n-1} E^{(n)}(\Delta_{l_n}) =$$
$$= \sum_{m=1}^{\infty} \sum_{l_1, \ldots, l_n = 1}^{\nu} \ldots = \sum_{m=1}^{\infty} \tilde{g}_{m,1}(A_1) T_1 \tilde{g}_{m,2}(A_2) \ldots T_{n-1} \tilde{g}_{m,n}(A_n).$$

It is obviously possible to interchange the order of summation because of inequality (2.7). To prove inequality (2.8), it now remains for us to note that we have $|\tilde{g}_{m,j}(\lambda)| \leqslant M_{m,j}$, and again make use of inequality (2.7).

For a given $\varepsilon > 0$, let us find a number m_0 such that

$$\sum_{m=m_0+1}^{\infty} M_{m,1} \ldots M_{m,n} < \varepsilon,$$

and let us consider the function

$$\psi(\lambda_1, \ldots, \lambda_n) = \sum_{m=1}^{m_0} \varphi_m(\lambda_1, \ldots, \lambda_n).$$

For any integral sum (2.4) we have the inequality

$$\left\| S - \sum_{m=1}^{\infty} J\varphi_m \right\| \leqslant \| S(\psi; T_1, \ldots, T_{n-1}; \Xi; \xi_1, \ldots, \xi_\nu) - J\psi \| +$$
$$+ \left\| S(\varphi - \psi; T_1, \ldots, T_{n-1}; \Xi; \xi_1, \ldots, \xi_\nu) - \sum_{m=m_0+1}^{\infty} J\varphi_m \right\| \leqslant \| S(\psi; T_1, \ldots, T_{n-1}; \Xi; \xi_1, \ldots, \xi_\nu) - J\psi \| + 2\varepsilon.$$

The first term can be made arbitrarily small because of the smallness of $h(\Xi)$. The validity of the lemma follows from this.

Corollary. The sum $\sum_{m=1}^{\infty} J\varphi_m$ is independent of the representation of the function φ in the form of series (2.6) satisfying condition (2.7), but is only governed by the function itself.

The above allows us to introduce the following definition.

Definition. Let the function φ be represented in the form of series (2.6) satisfying condition (2.7) and let all functions $g_{m,j}$, with $j = 1,\ldots, n$ and $m = 1, 2,\ldots$, be continuous. Integral (1.1) will be taken to mean the operator

$$Q = J(\varphi;\ T_1,\ \ldots,\ T_{n-1}) = \sum_{m=1}^{\infty} J(\varphi_m;\ T_1,\ \ldots,\ T_{n-1}). \tag{2.9}$$

Let **F** denote the class of functions satisfying this definition. It follows from the arguments given above that if the operators $T_1,\ldots, T_{n-1}$ are bounded and we have $\varphi \in \mathbf{F}$, then the operator (2.9) is bounded and its norm satisfies the inequality

$$\| J(\varphi;\ T_1,\ \ldots,\ T_{n-1}) \| \leqslant M \| T_1 \| \ldots \| T_{n-1} \|.$$

If some of the operators T_j, with $j = 1,\ldots, n-1$, possess better properties, then the properties of the operator (2.9) improve. We will formulate the properties of operators T_j in terms of their membership of classes $\mathbf{S}_p$, where $1 \le p \le \infty$. (See [8] concerning the classes $\mathbf{S}_p$.)

THEOREM 1. Let T_j belong to $\mathbf{S}_{p_j}$ (with $1 \le p_j \le \infty$, $j = 1,\ldots, n-1$) and let

$$p = \max\left(1;\ (p_1^{-1} + \ldots + p_{n-1}^{-1})^{-1}\right). \tag{2.10}$$

Then, for any function φ belonging to **F**, the series $\sum_{m=1}^{\infty} J(\varphi_m;\ T_1,\ \ldots,\ T_{n-1})$ converges in the form of space $\mathbf{S}_p$ and we have

$$\| J(\varphi;\ T_1,\ \ldots,\ T_{n-1}) \| \leqslant M \| T_1 \|_{p_1} \ldots \| T_{n-1} \|_{p_{n-1}}. \tag{2.11}$$

Proof. It is sufficient to make use of the inequality (see [8], Section 7, Chapter III)

$$\| J(\varphi_m;\ T_1,\ \ldots,\ T_{n-1}) \|_p \leqslant M_{m,1} \ldots M_{m,n} \| T_1 \|_{p_1} \ldots \| T_{n-1} \|_{p_{n-1}},$$

which, in view of (2.7), leads to the assertion of the theorem.

Remark. A result analogous to Theorem 1 also holds when some of the operators T_j are merely bounded. The corresponding p_j in formula (2.10) should then be replaced by ∞ and the factors $\| T_j \|_p$ in inequality (2.11) should be replaced by the usual operator norm $\| T_j \|$.

§ 3. The Fundamental Theorem

We will be interested in n-tuple integrals of functions which are difference relations of order $n-1$ for a function of one variable. Let us now summarize the information concerning difference relations that will be needed below.

Let the function f be defined on the interval $[a, b]$ and let it be differentiable a sufficient number of times. Its first-order difference relation is defined to be

$$\varphi(\lambda_1,\ \lambda_2) = \begin{cases} \dfrac{f(\lambda_2) - f(\lambda_1)}{\lambda_2 - \lambda_1} & (\lambda_2 \neq \lambda_1); \\ f'(\lambda_1) & (\lambda_2 = \lambda_1). \end{cases}$$

Let us agree to denote a first-order difference relation by $\Delta^{[1]}_{\lambda_1, \lambda_2} f$ or, more simply, by $f^{[1]}(\lambda_1, \lambda_2)$. Difference relations of higher order are defined inductively as follows:

$$f^{[k]}(\lambda_1, \ldots, \lambda_{k+1}) = \Delta^{[1]}_{\lambda_k, \lambda_{k+1}} f^{[k-1]}(\lambda_1, \ldots, \lambda_{k-1}).$$

It follows from this that, in particular, we have

$$f^{[k]}(\lambda, \ldots, \lambda) = \frac{f^{(k)}(\lambda)}{k!}.$$

Let us note two simple formulas that will be required below

1°. $\Delta^{[1]}_{\lambda_1, \lambda_2}(u \cdot v) = u(\lambda_1) v^{[1]}(\lambda_1, \lambda_2) + u^{[1]}(\lambda_1, \lambda_2) v(\lambda_2).$ (3.1)

2°. If $f(\lambda) = \lambda^s$ (with s a natural number), then for $k \le s$, we have

$$f^{[k]}(\lambda_1, \ldots, \lambda_{k+1}) = \sum_{\substack{s_1 > 0, \ldots, s_{k+1} > 0, \\ s_1 + \ldots + s_{k+1} = s-k}} \lambda_1^{s_1} \ldots \lambda_{k+1}^{s_{k+1}}. \tag{3.2}$$

The number of terms in the sum (3.2) is equal to $\binom{s}{k} = \frac{s(s-1)\ldots(s-k+1)}{k!}$.

The principal result of the present work is the following theorem.

THEOREM 2. Let each of the expansions of unity $E^{(j)}_\lambda$ with $j = 1, \ldots, n$ be constant outside the interval $[a, b]$, let the function f on this interval belong to class $C^{n-1+\varepsilon}$, where ε is greater than zero, and let $T_1, \ldots, T_{n-1}$ be bounded operators. Then, the integral $J(f^{[n-1]}; T_1, \ldots, T_{n-1})$ exists and its norm satisfies the inequality

$$\| J(f^{[n-1]}; T_1, \ldots, T_{n-1}) \| \leqslant c_{n,\varepsilon} \cdot \| f \|_{C^{n-1+\varepsilon}} \cdot \| T_1 \| \ldots \| T_{n-1} \|.$$

If, in addition, we have $T_j \in \mathbf{S}_{p_j}$, and $1 \leqslant p_j \leqslant \infty$ with $j = 1, \ldots, n-1$, then the operator $J(f^{[n-1]}; T_1, \ldots, T_{n-1})$ belongs to class $\mathbf{S}_p$, where p is defined by (2.10), and

$$\| J(f^{[n-1]}; T_1, \ldots, T_{n-1}) \|_p \leqslant c_{n,\varepsilon} \cdot \| f \|_{C^{n-1+\varepsilon}} \cdot \| T_1 \|_{p_1} \ldots \| T_{n-1} \|_{p_{n-1}}.$$

It follows from the results of Section 2 that it is sufficient to establish that $f^{[n-1]}$ belongs to $\mathbf{F}$. The proof of this inclusion wil require a number of tedious derivations; we begin by giving a number of preliminary remarks.

It is clear that we can set $[a, b] = [-\pi, \pi]$ and assume that the function f is periodic and belongs to class $C^{n-1+\varepsilon}$ on the whole axis. Let us expand f in a Fourier series

$$f(\lambda) = \sum_{m=-\infty}^{\infty} c_m e^{im\lambda}.$$

Let us introduce the functions

$$f_+(\lambda) = \sum_{m=1}^{\infty} c_m e^{im\lambda}, \quad f_-(\lambda) = \sum_{m=1}^{\infty} c_{-m} e^{-im\lambda}.$$

According to Privalov's theorem (for example, see [9]), the functions f_+ and f_- also belong to class $C^{n-1+\varepsilon}$ (we assume that $\varepsilon < 1$). This immediately allows us to restrict our investigation to functions of the form

$$f(\lambda)=\sum_{m=1}^{\infty} c_m e^{im\lambda}. \tag{3.3}$$

Let us introduce the quantity $\sigma_k(\lambda)$ defined by

$$\sigma_k(\lambda)=\sum_{m=k}^{\infty} c_m e^{i(m-k)\lambda}; \tag{3.4}$$

thus, the function $\sigma_k(\lambda)$ represents to within sign the remainder obtained when the function f is approximated by the first $k-1$ terms of its Fourier series. The Lebesgue theorem (see [9]) yields the inequality

$$|\sigma_k(\lambda)| \leqslant Ck^{-(n-1+\varepsilon)} \ln(1+k)\cdot \|f\|_{C^{n-1+\varepsilon}}. \tag{3.5}$$

The other auxiliary propositions will be formulated as two lemmas and corollaries.

LEMMA 2. *Let the function* $w(z)$ *be regular in the disc* $|z|<r$ *and expandable in a power series which converges absolutely in the closed disc* $|z| \leq r$,

$$w(z)=\sum_{k=0}^{\infty} a_k z^k.$$

Then, for arbitrary points z_1 *and* z_2, *with* $|z_1| \leq r$ *and* $|z_2| < r$, *we have*

$$w^{[1]}(z_1, z_2)=\sum_{k=1}^{\infty} z_2^{k-1} w_k(z_1), \tag{3.6}$$

where

$$w_k(z)=\sum_{m=k}^{\infty} a_m z^{m-k}. \tag{3.7}$$

Proof. Relation (3.6) formally follows from the following derivation:

$$\frac{w(z_2)-w(z_1)}{z_2-z_1}=\sum_{m=1}^{\infty} a_m \frac{z_2^m-z_1^m}{z_2-z_1}=\sum_{m=1}^{\infty} a_m \sum_{k=1}^{m} z_1^{m-k} z_2^{k-1}=\sum_{k=1}^{\infty} z_2^{k-1} \sum_{m=k}^{\infty} a_m z_1^{m-k}.$$

However, we should justify the change in the order of summations in the double series and to do this, we have only to prove that it is absolutely convergent. If $|z_2|=|z_1|$ (less than r), then we have

$$\sum_{m=1}^{\infty}\sum_{k=1}^{m} |a_m z_1^{m-k} z_2^{k-1}|=\sum_{m=1}^{\infty} m|a_m||z_1|^{m-1},$$

and the absolute convergence of the double series follows from the theory of power series. On the other hand, if we have $|z_2| \neq |z_1|$, then, assuming for definiteness that $|z_2|<|z_1|$, we find that

$$\sum_{m=1}^{\infty}\sum_{k=1}^{m} |a_m z_1^{m-k} z_2^{k-1}|=\sum_{m=1}^{\infty} |a_m| \frac{|z_1|^m-|z_2|^m}{|z_1|-|z_2|} \leqslant \frac{1}{|z_1|-|z_2|}\sum_{m=1}^{\infty} |a_m|\cdot|z_1|^m;$$

the last series converges when $|z_1| \leq r$ because of the conditions of the lemma. The lemma has been proved.

Corollary. Let us assume that $u(\lambda) = e^{i\lambda}$, where $-\infty < \lambda < \infty$. We have

$$u^{[1]}(\lambda_1, \lambda_2) = \sum_{k=1}^{\infty} \lambda_2^{k-1} u_k(\lambda_1), \tag{3.8}$$

the function u_k with $k = 1, 2, \ldots$ satisfying the inequality

$$\sup_{-\infty<\lambda<\infty} |u_k(\lambda)| \leqslant \frac{1}{k!}. \tag{3.9}$$

Indeed, relation (3.8) is a particular case of equality (3.6). In addition, we have

$$u_k(\lambda) = \lambda^{-k} \sum_{m=k}^{\infty} \frac{(i\lambda)^m}{m!} = \lambda^{-k} \left[e^{i\lambda} - \sum_{m=0}^{k-1} \frac{(i\lambda)^m}{m!} \right].$$

It follows from this that

$$|u_k(\lambda)| \leqslant |\lambda|^{-k} \cdot \frac{|\lambda|^k}{k!} \cdot \max_{\lambda} |(e^{i\lambda})^{(k)}| = \frac{1}{k!},$$

which was to be proved.

LEMMA 3. Let f be a periodic function of class $C^{1+\varepsilon}$, which can be expanded in a Fourier series of the form of (3.3). Then, for $\lambda_1, \lambda_2 \in [-\pi, \pi]$ we have

$$\frac{f(\lambda_2) - f(\lambda_1)}{e^{i\lambda_2} - e^{i\lambda_1}} = \sum_{k=1}^{\infty} e^{i(k-1)\lambda_2} \sigma_k(\lambda_1), \tag{3.10}$$

where σ_k is defined by (3.4).

Proof. Let us consider the function

$$w(z) = \sum_{m=1}^{\infty} c_m z^m.$$

In view of the conditions of the lemma, this series converges absolutely in the disc $|z| \leq 1$ and we have $w(e^{i\lambda}) = f(\lambda)$. We find from Lemma 1 that for points $z_1 = e^{i\lambda_1}$, $z_2 = \rho e^{i\lambda_2}$ (where $\rho < 1$), we have

$$\frac{w(z_2) - w(z_1)}{z_2 - z_1} = \sum_{k=1}^{\infty} \rho^{k-1} e^{i(k-1)\lambda_2} w_k(e^{i\lambda_1}). \tag{3.11}$$

In view of inequality (3.5), the series (3.11) is dominated by the convergent series $\sum_{k=1}^{\infty} Ck^{-(1+\varepsilon)} \ln(1+k) \cdot \|f\|_{C^{1+\varepsilon}}$ when $\rho \leq 1$. This allows us to perform the limiting transition $\rho \to 1-0$ which then leads to equality (3.10).

Corollary. Under the conditions of Lemma 3, we have

$$f^{[1]}(\lambda_1, \lambda_2) = \sum_{k=1}^{\infty} \sum_{l=1}^{\infty} \lambda_2^{k-1} e^{i(l-1)\lambda_2} u_k(\lambda_1) \sigma_l(\lambda_1). \tag{3.12}$$

Indeed, it is sufficient to compare formulas (3.8) and (3.10).

In particular, for functions $v_s(\lambda) = e^{is\lambda}$, where s is a natural number, we have

$$v_s^{[1]}(\lambda_1, \lambda_2) = \sum_{k=1}^{\infty} \sum_{l=1}^{s} \lambda_2^{k-1} e^{i(l-1)\lambda_2} u_k(\lambda_1) e^{i(s-l)\lambda_1}. \tag{3.13}$$

<u>Proof of Theorem 2.</u> Let us introduce the functions

$$F_\nu(\lambda_1, \ldots, \lambda_{\nu+1}) = \sum_{k_1, \ldots, k_\nu = 1}^{\infty} \sum_{1 \leq l_\nu < \cdots < l_1 < \infty} u_{k_1}(\lambda_1) \ldots u_{k_\nu}(\lambda_\nu) \times$$
$$\times \sigma_{l_1}(\lambda_1) e^{i(l_1 - l_2 - 1)\lambda_2} \ldots e^{i(l_{\nu-1} - l_\nu - 1)\lambda_\nu} e^{i(l_\nu - 1)\lambda_{\nu+1}} \lambda_{\nu+1}^{k_1 + \cdots + k_\nu - \nu} \tag{3.14}$$
$$(\nu = 1, \ldots, n-1).$$

Comparing this expression with (3.12), we find that

$$F_1(\lambda_1, \lambda_2) = f^{[1]}(\lambda_1, \lambda_2). \tag{3.15}$$

Subsequent derivations will show that the function F_ν represents the "principal term" of expression $f^{[\nu]}$.

Estimating the maximum absolute value of each term by means of inequalities (3.5) and (3.9), we find that the series (3.14) is dominated in the cube $\lambda_j \in [-\pi, \pi]$ (where j = 1,..., ν + 1) by the series

$$\sum_{k_1, \ldots, k_\nu = 1}^{\infty} \sum_{1 \leq l_\nu < \cdots < l_1 < \infty} C \frac{\ln(1 + l_1)}{l_1^{n-1+\varepsilon}} \|f\|_{C^{n-1+\varepsilon}} \cdot \frac{\pi^{k_1 + \cdots + k_\nu - \nu}}{k_1! \ldots k_\nu!} = C\left(\frac{e^\pi - 1}{\pi}\right)^\nu \cdot \sum_{l=1}^{\infty} \binom{l-1}{\nu-1} \frac{\ln(1+l)}{l^{n-1+\varepsilon}} \cdot \|f\|_{C^{n-1+\varepsilon}}.$$

The last series converges when we have $\nu < n$. We have thus established that we have $F_\nu \in \mathbf{F}$ (where ν = 1,..., n − 1).

Let us now proceed to the calculation of the difference relations for the function F_ν. Making use of formula (3.1), we find that

$$\Delta^{[1]}_{\lambda_\nu, \lambda_{\nu+1}} F_{\nu-1}(\lambda_1, \ldots, \lambda_{\nu-1}, \cdot) = F_\nu(\lambda_1, \ldots, \lambda_{\nu+1}) + \sum_{k_1, \ldots, k_{\nu-1} = 1}^{\infty} \sum_{1 \leq l_{\nu-1} < \cdots < l_1 < \infty} u_{k_1}(\lambda_1) \ldots u_{k_{\nu-1}}(\lambda_{\nu-1}) \sigma_{l_1}(\lambda_1) \times$$
$$\times e^{i(l_1 - l_2 - 1)\lambda_2} \ldots e^{i(l_{\nu-2} - l_{\nu-1} - 1)\lambda_{\nu-1}} e^{i(l_{\nu-1} - 1)\lambda_\nu} \Delta^{[1]}_{\lambda_\nu, \lambda_{\nu+1}} \{\lambda^{k_1 + \cdots + k_{\nu-1} - \nu + 1}\}.$$

It follows from this that we have

$$\Delta^{[n-\nu]}_{\lambda_\nu, \ldots, \lambda_n} F_{\nu-1}(\lambda_1, \ldots, \lambda_{\nu-1}, \cdot) = \Delta^{[n-\nu-1]}_{\lambda_{\nu+1}, \ldots, \lambda_n} F_\nu(\lambda_1, \ldots, \lambda_\nu, \cdot) +$$
$$+ \sum_{k_1, \ldots, k_{\nu-1} = 1}^{\infty} \sum_{1 \leq l_{\nu-1} < \cdots < l_1 < \infty} u_{k_1}(\lambda_1) \ldots u_{k_{\nu-1}}(\lambda_{\nu-1}) \sigma_{l_1}(\lambda_1) \times$$
$$\times e^{i(l_1 - l_2 - 1)\lambda_2} \ldots e^{i(l_{\nu-2} - l_{\nu-1} - 1)\lambda_{\nu-1}} e^{i(l_{\nu-1} - 1)\lambda_\nu} \Delta^{[n-\nu]}_{\lambda_\nu, \ldots, \lambda_n} \{\lambda^{k_1 + \cdots + k_{\nu-1} - \nu + 1}\} \tag{3.16}$$
$$(\nu = 2, \ldots, n-1).$$

Let us denote the second term on the right-hand side of the last equality by $G_{\nu-1}(\lambda_1, \ldots, \lambda_n)$. If we take formula (3.2) into account, then it is clear that $G_{\nu-1}$ represents the sum of a series of the form of (2.6). Estimating the maximum absolute value of each term, we find that within the cube $\lambda_j \in [-\pi, \pi]$ (where j = 1,..., n) this series is dominated by

$$\sum_{k_1, \ldots, k_{\nu-1} = 1}^{\infty} \sum_{1 \leq l_{\nu-1} < \cdots < l_1 < \infty} C \frac{\ln(1 + l_1)}{l_1^{n-1+\varepsilon}} \|f\|_{C^{n-1+\varepsilon}} \binom{k_1 + \cdots + k_{\nu-1} - \nu + 1}{n - \nu} \cdot \frac{\pi^{k_1 + \cdots + k_{\nu} - n + 1}}{k_1! \ldots k_{\nu-1}!}.$$

The last series converges; we therefore find that $G_{\nu-1} \in \mathbf{F}$ (where ν = 2,..., n − 1).

Comparing expressions (3.15) and (3.16), we find that

$$f^{[n-1]} = F_{n-1} + G_1 + \ldots + G_{n-2}.$$

It now immediately follows that $f^{[n-1]}$ belongs to **F**. The theorem has been proved.

Remark. The existence of the integral of the function $\Delta^{[n-1]}f$ has been proved in [1] under more stringent conditions on the function f. It was assumed in [1] that f belongs to C^{2n-2}.

In the same way as in [1], the theorem concerning integrals of the type under consideration can be used to derive some results on the differentiation of functions of self-adjoint bounded operators depending on a parameter. These results are based on a formula obtained in [1] and given below.

Let T(t) be an operator function depending on the parameter t, the values of T(t) being self-adjoint bounded operators in Hilbert space. We will restrict ourselves to the simplest case when this function is linear, namely,

$$T(t) = T_0 + tT.$$

Let E_λ denote the expansion of unity for the operator T_0; let us assume that it is constant outside the interval $[a, b]$. Let the function f defined on $[a, b]$ belong to class $C^{n+\varepsilon}$. Then, the operator function $f(T_0 + tT)$ is differentiable n times and we have

$$\left[\frac{d^n}{dt^n} f(T_0 + tT)\right]_{t=0} = n! \underbrace{\int \ldots \int}_{n+1} f^{[n]}(\lambda_1, \ldots, \lambda_{n+1}) \, dE_{\lambda_1} T dE_{\lambda_2} \ldots T dE_{\lambda_{n+1}}.$$

(This formula was obtained in [1] under more stringent restrictions on the function f.)

It should also be noted that, as follows from Theorem 2, the derivative $f^{(n)}(T_0 + tT)$ for for $T \in S_p$ (with $1 \leq p \leq \infty$) is an operator of class S_q, where $q = \max\left(\frac{p}{n}, 1\right)$.

Literature Cited

1. Yu. L. Daletskii and S. G. Krein, "Integration and differentiation of functions of Hermitian operators with applications in perturbation theory," in: Transactions of the Voronezh Seminar on Functional Analysis [in Russian], Vol. 1, Voronezh (1956), pp. 81-205.
2. M. Sh. Birman and M. Z. Solomyak, "On Stieltjes double-integral operators," Dokl. Akad. Nauk SSSR, Vol. 165, No. 6, pp. 1223-1226 (1965).
3. M. Sh. Birman and M. Z. Solomyak, "Stieltjes double-integral operators," in: Topics in Mathematical Physics, Vol. 1, M. Sh. Birman (editor), Consultants Bureau, New York (1967), pp. 25-54.
4. M. Sh. Birman and M. Z. Solomyak, "Stieltjes double-integral operators and the problem of factors," Dokl. Akad. Nauk SSSR, Vol. 171, No. 6 (1966).
5. M. Sh. Birman and M. Z. Solomyak, "Stieltjes double-integral operators. II," in: Topics in Mathematical Physics, Vol. 2, M. Sh. Birman (editor), Consultants Bureau, New York (1968), pp. 19-46.
6. B. S. Pavlov, "On multidimensional integral operators," Present Volume, p.
7. K. Töllner, "Some properties of transformers defined by double-integral operators," Present Volume, p. 81.
8. I. Ts. Gokhberg and M. G. Krein, Introduction to the Theory of Nonself-Adjoint Operators [in Russian], Izd. "Nauka," Moscow (1966).
9. A. Zigmund, Trigonometric Series, Vols. I and II [Russian translation], Izd. "Mir," Moscow (1965).

ON THE PROPAGATION OF LOVE WAVES ALONG THE SURFACE OF AN INHOMOGENEOUS ELASTIC BODY OF ARBITRARY SHAPE

V. M. Babich and N. Ya. Kirpichnikova

The propagation of Love waves along the surface of an inhomogeneous elastic body of arbitrary shape is studied in the present article. The parabolic-equation method is used to determine the dependence of Love waves on the frequency and position. Formulas for the phase velocities and penetration depths of these waves are obtained.

Some of the procedures described in [1, 2] are used in this article.

Let us consider an inhomogeneous elastic body bounded by a sufficiently smooth surface S. In addition to the Cartesian coordinate system (x, y, z), let us introduce a curvillinear coordinate system (τ, α, ν) with radial coordinates. The parameters τ, α, and ν have the following meaning.

Let us assume that a wave propagates along the surface S with a velocity b(x, y, z). Let us choose the coordinate lines (α) such that at any time they coincide with the wave front.

Let the coordinate line (α) coincide with the wave front at time t = 0. Let us draw through any point of this line, orthogonally to it on S, the extremals of the integral

$$\int_{M_0}^{M} \frac{ds}{b(x, y, z)}. \tag{1}$$

The position of the wave front at time $t = t_0$ is the locus of points M

$$t_0 = \int_{M_0}^{M} \frac{ds}{b(x, y, z)},$$

the integration being performed over the extremal of integral (1) on S. Let us choose the coordinate lines (τ) to be the extremals of integral (1) such that

$$\tau(x, y, z) = t = \int_{M_0}^{M} \frac{ds}{b(x, y, z)}. \tag{2}$$

The parameter ν is the distance from the surface S and points within the elastic body have $\nu \geq 0$.

Any radius vector $\vec{x} = (x, y, z)$ of an arbitrary point near the surface S can be represented as

$$\vec{x} = \vec{r}(\tau, \alpha) + \nu \vec{n}(\tau, \alpha), \tag{3}$$

where $\vec{r} = \vec{r}(\tau, \alpha)$ is the parametric equation of the surface, and $\vec{n} = \vec{n}(\tau, \alpha)$ is the inward-pointing normal to the surface S. Thus, τ, α, and ν can be considered as curvilinear coordinates defined in the neighborhood of the surface. Let $g_{\tau\tau}$, $g_{\tau\alpha}$, and $g_{\alpha\alpha}$ denote the coefficients of the first quadratic form of surface S and let $b_{\tau\tau}$, $b_{\tau\alpha}$, and $b_{\alpha\alpha}$ denote the coefficients of the second quadratic form.

The square of an element of arc length can be represented as

$$ds^2 = G_{ij}\, dq^i\, dq^j, \quad i, j = 1, 2, 3. \tag{4}$$

Here $\tau = q^1$, $\alpha = q^2$, and $\nu = q^3$; the summation convention is adopted for repeated indices; G_{ij} is the metric tensor in the space of the coordinates q^1, q^2, and q^3. To within terms of order ν^2, the coefficients of this tensor are given by

$$\left.\begin{aligned} G_{\tau\tau} &= g_{\tau\tau} - 2\nu b_{\tau\tau} + 0(\nu^2), & G_{\tau\nu} &= 0, \\ G_{\tau\alpha} &= \quad\ \ - 2\nu b_{\tau\alpha} + 0(\nu^2), & G_{\alpha\nu} &= 0, \\ G_{\alpha\alpha} &= g_{\alpha\alpha} - 2\nu b_{\alpha\alpha} + 0(\nu^2), & G_{\nu\nu} &= 1. \end{aligned}\right\} \tag{5}$$

Since we have

$$G_{ij}G^{jk} = \delta_i^k \ (\delta_i^k - \text{is the Kronecker symbol}), \tag{6}$$

it is not difficult to calculate G^{jk} to the same accuracy in ν. Let the medium under consideration have Lamé parameters λ and μ and density ρ which are functions of the coordinates τ, α, and ν.

We will assume that the condition

$$K + \frac{\partial \ln b}{\partial \nu} > 0, \tag{7}$$

where K is the curvature of the normal section of S along a ray, is satisfied on the surface S.

Let us assume that the surface S of the body is free from stresses and that there are no bulk forces. We will restrict ourselves to small displacements of the elastic medium. Let the coordinates of a point before displacement be (τ, α, ν) and the coordinates of the same point after displacement be $(\tau + \varphi^\tau, \alpha + \varphi^\alpha, \nu + \varphi^\nu)$. The contravariant components of the displacement vector in the coordinate system τ, α, ν will be

$$(\varphi^\tau, \varphi^\alpha, \varphi^\nu),$$

where φ^i is a small displacement along the i-th coordinate.

If a small displacement vector is defined at every point of the medium, then we consider that the displacement of the whole medium is given.

The equation of motion of an inhomogeneous elastic medium is

$$\sigma_{ij} G^{ii'} G^{jj'} \frac{\partial G_{i'j'}}{\partial q^s} \sqrt{g} - 2 \frac{\partial}{\partial q^j} (\sigma_{sj'} \cdot G^{jj'} \sqrt{g}) + 2\rho G_{sj} \sqrt{g} \frac{\partial^2 \varphi^j}{\partial t^2} = 0, \tag{8}$$

where i, i', j, j' and s vary from 1 to 3, $g = \det \| G_{ij} \|$; ω is the frequency, and σ_{ij} the stress tensor.

It can be shown that in our case, Hooke's law can be expressed as

$$\sigma_{ij} = \lambda \frac{1}{\sqrt{g}} \frac{\partial}{\partial q^s} (\varphi^s \sqrt{g}) \delta_i^k G_{kj} + \mu \left[\frac{\partial G_{ij}}{\partial q^s} \varphi^s + G_{sj} \frac{\partial \varphi^s}{\partial q^i} + G_{js} \frac{\partial \varphi^s}{\partial q^i} \right]. \tag{9}$$

The unknown vector $\vec{u}$ satisfying Eqs. (9) must be such that there are no stresses on the boundary, i.e.,

$$\sigma_{i\nu} \big|_{\nu=0} = 0 \quad (i = \tau,\ \alpha,\ \nu). \tag{10}$$

Let the displacement vector be

$$\vec{u} = \exp\left[-i\omega(t-\tau)\right] \cdot \bar{V}(\tau,\ \alpha,\ \nu,\ \omega), \tag{11}$$

where $\bar{V}$ is a vector with the coordinates

$$\bar{V} = (V^{(\tau)},\ V^{(\alpha)},\ V^{(\nu)}). \tag{12}$$

Let us substitute (11) into Eq. (8) and the boundary conditions (10). Considering only the principal terms of the equation with respect to parameter ω ($\omega \to \infty$) of order $O(\omega^{2/3})$ and assuming by analogy with [1] that a differentiation of $V^{(i)}$, where $i = \tau$, α, and ν, with respect to τ, α, and ν is equivalent in order of magnitude to a multiplication by $\omega^{1/3}$, $\omega^{1/3}$, and $\omega^{2/3}$, respectively, we obtain equations for $V^{(\tau)}$, $V^{(\alpha)}$, and $V^{(\nu)}$ and the corresponding boundary conditions.

Let us take

$$V^{(\tau)} = \omega^{-2/3} V_1^{(\tau)} + \omega^{-1} V_2^{(\tau)}, \tag{13}$$

where

$$V_1^{(\tau)} = i \frac{\dfrac{\partial V^{(\alpha)}}{\partial \alpha} + \dfrac{\partial V^{(\nu)}}{\partial \nu}}{\omega^{1/3}};$$

$$V_2^{(\tau)} = i \frac{(\lambda+\mu)\dfrac{\partial}{\partial \alpha} \ln\sqrt{g} + \mu \dfrac{\partial}{\partial \alpha} \ln g_{\tau\tau} + \mu \dfrac{\partial}{\partial \alpha} \ln \mu}{\lambda + \mu} V^{(\alpha)}.$$

The function $V^{(\alpha)}$ is such that

$$2\sqrt{g}\,\mu\omega^2 G_{\alpha\alpha}\left(\frac{\rho}{\mu} - \frac{1}{g_{\tau\tau}} - 2\frac{b_{\tau\tau}}{g_{\tau\tau}^2}\nu\right) V^{(\alpha)} +$$

$$+ 4i\omega\sqrt{g}\,\mu G_{\alpha\alpha} G^{\tau\tau} \frac{\partial V^{(\alpha)}}{\partial \tau} + 2\sqrt{g}\,\mu G_{\alpha\alpha} \frac{\partial^2 V^{(\alpha)}}{\partial \nu^2} + 2\frac{\partial}{\partial \tau}\left(\mu G_{\alpha\alpha} \sqrt{g}\, G^{\tau\tau}\right) i\omega V^{(\alpha)} + O(\omega^{2/3}) = 0, \tag{14}$$

$$\frac{\partial V^{(\alpha)}}{\partial \nu}\bigg|_{\nu=0} = 0. \tag{15}$$

Let us also assume that

$$V^{(\nu)} = \omega^{-1/3} e^{\omega^{1/3}\varphi(\tau,\alpha)} V_1^{(\nu)}, \tag{16}$$

where $\varphi(\tau,\ \alpha)$ is a function defined below by formula (24) and $V_1^{(\nu)}$ is the solution of the following Sturm–Liouville problem:

$$\frac{\partial^2 V_1^{(\nu)}}{\partial \xi^2} - (\xi A + B) V_1^{(\nu)} = F(\xi), \tag{17}$$

$$V_1^{(\nu)}|_{\xi=0}=0,\ V_1^{(\nu)}\underset{\xi\to\infty}{\longrightarrow}0.^{*} \tag{18}$$

Here, A, B, F, and ξ are given by

$$\begin{aligned} A &= 2\frac{b_{\tau\tau}}{g_{\tau\tau}^2}-\frac{\partial}{\partial\nu}\frac{1}{b^2}, \\ B &= -2i\frac{1}{g_{\tau\tau}}\cdot\frac{\partial\varphi_{(\tau,\alpha)}}{\partial\tau}, \\ F &= -2i\frac{b_{\tau\alpha}}{g_{\tau\tau}}\cdot e^{-\omega^{1/3}\varphi(\tau,\alpha)}V^{(\alpha)}, \\ \xi &= \omega^{2/3}\nu. \end{aligned} \tag{19}$$

Conditions (13)-(19) ensure that Eqs. (8) are satisfied [the right-hand sides are of order $O(\omega^{1/3})$]. The substitution into the boundary conditions yields

$$\sigma_{\tau\nu}|_{\nu=0}=O(\omega^{1/3}),\quad \sigma_{\alpha\nu}|_{\nu=0}=O(\omega^{1/3}),\quad \sigma_{\nu\nu}|_{\nu=0}=O(\omega^{1/3}).$$

When $V^{(\alpha)}$ satisfies Eq. (14) and the boundary condition (15), $V^{(\alpha)}$ tends to 0 as $\nu\to\infty$, and $V^{(\alpha)}$ is not identically zero, the displacement vector $\vec{u}$ will be called a Love wave.

Equation (14) will be solved by procedures that are traditional for this type of problem.

Let us introduce the independent variable

$$\xi_1=\omega^{2/3}\cdot\nu\cdot\psi(\tau,\alpha) \tag{20}$$

and the function

$$W(\tau,\alpha,\xi_1)=e^{-\omega^{1/3}\varphi(\tau,\alpha)}\cdot V^{(\alpha)}(\tau,\alpha,\nu). \tag{21}$$

Then, writing down only terms of higher order than $O(\omega^{2/3})$ and cancelling $\omega^{4/3}$, we find that Eq. (14) becomes

$$\frac{\partial^2 W}{\partial\xi_1^2}+\left[\left(\frac{\partial}{\partial\nu}\frac{1}{b^2}-2\frac{b_{\tau\tau}}{g_{\tau\tau}^2}\right)\frac{1}{\psi^3(\tau,\alpha)}\cdot\xi_1 W+2i\frac{1}{g_{\tau\tau}}\cdot\frac{\frac{\partial}{\partial\tau}\varphi(\tau,\alpha)}{\psi^2(\tau,\alpha)}W\right]+\frac{i}{\omega^{1/3}g_{\tau\tau}\psi^2(\tau,\alpha)}\left[2\frac{\partial W}{\partial\tau}+\right.$$
$$\left.+2\xi_1\frac{\frac{\partial}{\partial\tau}\psi(\tau,\alpha)}{\psi(\tau,\alpha)}\frac{\partial W}{\partial\xi_1}+\frac{\partial}{\partial\tau}\ln\left(\mu\frac{g_{\alpha\alpha}}{g_{\tau\tau}}\sqrt{g}\right)W\right]+O(\omega^{-2/3})=0. \tag{22}$$

Let us choose $\psi(\tau,\alpha)$ and $\varphi(\tau,\alpha)$ as follows:

$$\psi(\tau,\alpha)=\sqrt[3]{2\frac{b_{\tau\tau}}{g_{\tau\tau}^2}-\frac{\partial}{\partial\nu}\frac{1}{b^2}}>0, \tag{23}$$

$$\varphi(\tau,\alpha)=-i\frac{\zeta}{2}\cdot\int_{M_0}^{M}\left(2\frac{b_{\tau\tau}}{g_{\tau\tau}^2}-\frac{\partial}{\partial\nu}\frac{1}{b^2}\right)^{2/3}g_{\tau\tau}\,d\tau, \tag{24}$$

where ζ is a constant which plays the part of a separation constant.

---

* Subsequent derivations show that $v'\left(\frac{B}{A^{2/3}}\right)=0$, so that we have $v\left(\frac{B}{A^{2/3}}\right)\neq 0$; because of this, problem (17)-(18) has a unique solution for any choice of $F(\xi)$.

After transformations analogous to those carried out in [1], we obtain

$$\varphi_j^{(\alpha)} = \exp[-i\omega(t-\tau)] \cdot \frac{1}{\sqrt{g_{\alpha\alpha}}} \cdot W_j^{(\alpha)},$$

$$W_j^{(\alpha)} = C_j \cdot \frac{1}{\sqrt[4]{g_{\alpha\alpha}}} \cdot \frac{\sqrt[6]{2 \cdot \frac{1}{b}\left(\frac{b_{\tau\tau}}{g_{\tau\tau}} + \frac{\partial}{\partial\nu}\ln b\right)}}{\left(\frac{1}{b}\right)^{1/3} \cdot \mu^{1/2}} \exp \omega^{1/3} \cdot \left\{-i\frac{\zeta_j}{2}\int_{M_0}^{M}\left(\frac{2b_{\tau\tau}}{g_{\tau\tau}^2} - \frac{\partial}{\partial\nu}\frac{1}{b^2}\right)^{2/3} g_{\tau\tau}\,d\tau\right\} \times$$

$$\times\, v\left[\omega^{2/3}\left(2\frac{b_{\tau\tau}}{g_{\tau\tau}^2} - \frac{\partial}{\partial\nu}\frac{1}{b^2}\right)^{1/3}(\nu - h_j)\right] \times [1 + 0(\omega^{-1/3})]. \tag{25}$$

Here, ζ_j are the roots of the equation $v'(-\zeta) = 0$, $\zeta = \zeta_j$ (with j = 1, 2, 3,...), $v(t) \sim \frac{1}{2} t^{-1/4} e^{-2/3 t^{3/2}}$ is the Airy integral [3], and c_j is a constant. The quantity

$$h_j = \omega^{-2/3}\left(2\frac{b_{\tau\tau}}{g_{\tau\tau}^2} - \frac{\partial}{\partial\nu}\frac{1}{b^2}\right)^{-1/3}\zeta_j \tag{26}$$

is the depth to which the j-th Love wave penetrates into the medium.

Let us introduce a Cartesian coordinate system x_1, x_2, x_3, the origin being placed at the point $M_0 \in S$; here x_1 is the tangent to the coordinate line (τ) at the point M_0, x_2 the tangent to α, and x_3 the tangent to ν.

In this coordinate system, the displacement vector will be

$$\vec{u} = (u_1,\ u_2,\ u_3), \tag{27}$$

where

$$\begin{aligned} u_1 &= \exp[-i\omega(t-\tau)] \cdot \sqrt{g_{\tau\tau}}\, V^{(\tau)}, \\ u_2 &= \exp[-i\omega(t-\tau)] \cdot \sqrt{g_{\alpha\alpha}}\, V^{(\alpha)} = \exp[-i\omega(t-\tau)] \cdot W^{(\alpha)}, \\ u_3 &= \exp[-i\omega(t-\tau)] \cdot V^{(\nu)}. \end{aligned} \tag{28}$$

The quantities $V^{(\tau)}$, $V^{(\nu)}$, and $W^{(\alpha)}$ are given by formulas (13), (16)-(19), and (25).

A comparison of expressions (27) and (28) for the displacement vector with formulas (3) and (24)-(26) of [1] shows that the latter follow from our formulas as a special case. The quantity $g_{\alpha\alpha}$ which characterizes the divergence of the ray field on the surface S is identically equal to unity under the conditions of the problem considered in [1].

Literature Cited

1. V. M. Babich and I. A. Molotkov, "On the propagation of Love waves in an elastic half-space which is inhomogeneous along two coordinates," Izv. AN SSSR, Ser. Fizika Zemli, No. 6 (1966).
2. I. V. Mukhina and I. A. Molotkov, "On the propagation of Rayleigh waves in an elastic half-space which is inhomogeneous along two coordinates," Izv. AN SSSR, Ser. Fizika Zemli, No. 4 (1967).
3. V. A. Fok, Tables of Airy Functions [in Russian] (1946).

THE DIRICHLET PROBLEM FOR TWO-DIMENSIONAL QUASI-LINEAR SECOND-ORDER ELLIPTIC EQUATIONS

N. M. Ivochkina

The present article is devoted to an investigation of the solubility of the Dirichlet problem in the class $C_{2,\alpha}(\overline{\Omega})$* for two-dimensional quasi-linear second-order elliptic equations of the form

$$L(u;\ u)=a_{ij}(x,\ u,\ u_x)\,u_{x_i x_j}=a(x,\ u,\ u_x),\dagger \tag{1}$$

in a bounded simply-connected domain Ω of two-dimensional Euclidean space E^2. The functions $a_{ij}(x, u, p)$ and $a(x, u, p)$ are defined on the set $\mathfrak{M}=\{x\in\overline{\Omega},\ |u|\leqslant M,\ |p|<\infty\}$, where M is a known constant. The condition that the operator L(u;) is elliptic can be written as

$$\nu(M)\,\xi_i^2\leqslant a_{ij}\xi_i\xi_j\leqslant\mu(M)\,\lambda(|p|)\,\xi_i^2, \tag{2}$$

where $\nu(M)$ and $\mu(M)$ are positive constants and $\lambda(|p|)$ is a positive nondecreasing function. In the following, we will also assume the condition

$$\left|\frac{a(x,\ u,\ p)}{1+p^2}\right|\leqslant K,\ \ K=\text{const}>0. \tag{3}$$

Restriction (3) guarantees that Eq. (1) belongs to the class L introduced by S. N. Bernshtein (see [2], p. 21). There are examples (see [1, 2]) which indicate that this is a natural restriction in the sense that if it is not satisfied, then the solution of the Dirichlet problem for Eq. (1) will depend on the concrete form of the boundary to domain Ω and on the boundary condition even in the uniformly elliptic case.

The solubility of the Dirichlet problem

$$\begin{gathered} L(u;\ u)=a(x,\ u,\ u_x),\\ u\,|_{\partial\Omega}=\psi(x), \end{gathered} \tag{4}$$

where $\partial\Omega$ is the boundary of domain Ω, will be derived on the basis of the Lehrer–Schauder fixed-point principle, the application of which requires *a priori* estimates of the solution of

* The notation for the spaces is the same as that used in [1].

† Unless otherwise qualified, repeated indices in the following imply a summation, the range of the indices being 1 and 2.

problem (4) in the $C_{2,\alpha}(\bar{\Omega})$ norm. It should be noted that as soon as we are able to estimate $\max_{\bar{\Omega}}|u_x|$, Eq. (1) no longer differs in its properties from uniformly elliptic equations $[\lambda(|p|) \leq \mu_1]$. Therefore, the main part of the present article consists in the calculation of estimates of $\max_{\bar{\Omega}}|u_x|$ for problem (4). The theorem of the existence of a solution of problem (4) in the class $C_{2,\alpha}(\bar{\Omega})$ will finally be given without proof as a consequence of these estimates and the well-known results of [1] on uniformly elliptic equations. In constructing the *a priori* estimates, we will make use of procedures originating with S. N. Bernshtein, as well as some results of [1].

Let us begin with estimates on the boundary $\partial\Omega$. Let us first of all assume that the domain Ω is convex (not necessarily strictly convex) and that $u(x)|_{\partial\Omega} = 0$. Let us give a well-known proposition.

LEMMA 1. Let u(x) be the solution of problem (4), let its first-order derivatives be differentiable at every point $x \in \Omega$, let $u|_{\partial\Omega} = 0$, and let Ω be convex. In addition, let us assume that conditions (2) and (3) are satisfied and that

$$\max_{\bar{\Omega}}|u(x)| \leqslant M, \tag{5}$$

then the upper bound to $\max_{\partial\Omega}|u_x|$ is a constant which depends on M, μ, ν, K, and Ω.

A proof of this lemma is contained, for example, in [1] (p. 394). The assumption made there that the equation is uniformly elliptic is irrelevant.

Remark. If we have $\partial\Omega \in C_2$, we can smoothly transform an arbitrary simply-connected bounded domain Ω into a convex domain. However, as can be easily verified, if we have $\frac{\lambda(|p|)}{1+|p|} > h(|p|)$, where $h(\tau)$ is any positive monotonically increasing function, then as a result of the transformation of Ω, Eq. (1) may no longer belong to class L and, therefore, nonconvex domains are only admissible when we have

$$\frac{\lambda(|p|)}{1+|p|} \leqslant K.$$

Let us now consider the case when we have $u(x)|_{\partial\Omega} = \psi(x) \not\equiv 0$. When we have $\psi(x) \in C_2(\partial\Omega)$ and $\frac{\lambda(|p|)}{1+p^2} \leqslant K$, it is easy to arrive at the situation considered in Lemma 1. It is sufficient to consider the function $v(x) = u(x) - \bar{\psi}(x)$, where $\bar{\psi}(x)$ is an arbitrary function which belongs to $C_2(\bar{\Omega})$ and which coincides with $\psi(x)$ on the boundary $\partial\Omega$. This class of equations, in particular, will contain the equation of minimal surfaces. This procedure will be unsuitable when we have $\frac{\lambda(|p|)}{1+p^2} > h(|p|)$, because condition (3) may now be violated.

However, for strictly convex domains, the result of the lemma remains valid for arbitrary $\psi(x)$ belonging to $C_2(\partial\Omega)$. Let us establish an auxiliary lemma in order to make this point clear.

LEMMA 2. Let us consider the problems

$$L(u;\ \psi_1) \equiv a_{ij}(x,\ u,\ u_x)\psi_{1x_ix_j} \geqslant 0,\ \psi_1|_{\partial\Omega} = \psi(x), \tag{6}$$

$$L(u;\ \psi_2) \leqslant 0,\ \psi_2|_{\partial\Omega} = \psi(x). \tag{6'}$$

Let us assume that the operator $L(u;)$ is elliptic, domain Ω is strictly convex, $\partial\Omega$ belongs to C_2, and $\psi(x)$ belongs to $C_2(\partial\Omega)$. Then, problems (6) and (6') always have solutions belonging to class $C_2(\bar{\Omega})$. It should be noted that, for example, these solutions may be convex functions.

Proof. We will construct a function $\psi_1(x)$ which is convex upwards and such that $\psi_1(x)|_{\partial\Omega} = \psi(x)$. It is clear that $\psi_1(x)$ will be a solution of problem (6'). Without loss of generality, we can assume that the coordinate origin is at the center of the circle of radius r inscribed in Ω.

Let the axis $\vec{x}_3$ be directed upwards, at right angles to this circle. Let us take on $\vec{x}_3$ a point x_3^0 such that $x_3^0 > \max \psi(x)$ and let us construct a cone with vertex at x_3^0 and base formed by the three-dimensional curve $l = \{x = (x_1, x_2) \in \partial\Omega, \; x_3 = \psi(x)\}$. In general, the section $\Omega(x_3^0)$ of this cone by the plane $x_3 = 0$ will not be a convex figure, but if x_3^0 tends to ∞, then at some finite time it will become convex when $x_3^0 = a$. This is guaranteed by the assumption that Ω is strictly convex, i.e., by the inequality $k > \delta > 0$, where k is the curvature of $\partial\Omega$ and δ a positive number. Let us find the value of a required for $\Omega(a)$ to be convex. Let us write the equation of $\partial\Omega$ in the form $r = r(\varphi)$, where r is the radius vector joining the coordinate origin to the point on $\partial\Omega$ and φ is a polar angle. We then have

$$k = \frac{r^2 + 2r'^2 - rr''}{(r^2 + r'^2)^{3/2}} \geqslant \delta,$$

from which it follows that $r^2 + 2r'^2 - rr'' \geq \delta r_0^3$. Let us take the equation of the boundary $\partial\Omega(x_3^0)$ of section $\Omega(x_3^0)$ to be $\rho = \rho(\varphi, x_3^0)$. We have to find the value of a such that inequality $\rho^2 + 2\rho'^2 - \rho\rho'' \geq 0$ holds for $x_3^0 = a$. It is obvious that we have

$$\rho(\varphi, x_3^0) = \frac{x_3^0 r(\varphi)}{x_3^0 - \psi(\varphi)}.$$

We then have

$$\rho^2 + 2\rho'^2 - \rho\rho'' = \left[\frac{r^2 + 2r'^2 - rr''}{(x_3^0 - \psi)^2} + \frac{(r^2)'\psi' - r^2\psi''}{(x_3^0 - \psi)^3}\right](x_3^0)^2$$

and if we set

$$a = \max_{\partial\Omega}\left\{\frac{|r^2\psi'' - (r^2)'\psi'|}{\delta r_0^3} + |\psi|\right\},$$

the section $\Omega(a)$ and the cone with vertex at the point $x_1 = x_2 = 0$ and $x_3 = a$ will both be convex figures. Let us now smoothly replace the apex of the cone by a smooth convex "cap" (in such a manner that the surface remains convex). This surface will be the required solution of problem (6). The function which is equal to $-\psi(x)$ on the boundary $\partial\Omega$, satisfies inequality (6), and multiplied by (−1) will be the solution of problem (6'). Let us finally note that we have

$$\max_{\partial\Omega}\left|\frac{\partial\psi_1}{\partial r}\right| \leqslant \frac{a + \max\limits_{\partial\Omega}|\psi|}{r_0}.$$

The lemma has been proved.

Let us return to the assertion that was stated before Lemma 2. If Ω is strictly convex and $\lambda(|p|)$ is arbitrary, then by considering the functions $v_i(x) = u(x) - \psi_i(x)$ (depending on whether $u_x|_{\partial\Omega}$ is bounded from above or below), we arrive at the inequalities

$$L(u;\ v_1) \geqslant a(x,\ (u+\psi_1),\ (u+\psi_1)_x), \tag{4'}$$

$$L(u;\ v_2) \leqslant a(x,\ (u+\psi_2),\ (u+\psi_2)_x), \tag{4''}$$

$$v_i|_{\partial\Omega} = 0.$$

Since in the proof of Lemma 1 for the bound on the gradient of the solution of problem (4) on the boundary $\partial\Omega$, for example, from above, we in fact only required that an inequality of the type of (4') be satisfied, we are able to use this procedure to estimate $\max\limits_{\partial\Omega}|u_x|$ for the case under consideration.

Let us prove two auxiliary lemmas before proceeding to inner estimates.

LEMMA 3. Let g(p) be a differentiable function defined in n-dimensional Euclidean space E^n and equal to a constant in the neighborhood of the point p = 0. Then, g(p) can be represented as

$$g(p) = n\mathfrak{g}(p) + \sum_{i=1}^{n} p_i \frac{\partial \mathfrak{g}}{\partial p_i},\quad p = (p_1,\ \ldots,\ p_n), \tag{7}$$

where

$$\mathfrak{g}(p) = \frac{1}{r^n}\int_0^r g(\tau,\ \varphi)\,\tau^{n-1}\,d\tau, \tag{8}$$

and $\{r,\ \varphi\}$ are spherical coordinates in E^n with $\varphi = (\varphi_1, \ldots, \varphi_{n-1})$.

This proposition is proved by a direct method

$$p_i \frac{\partial \mathfrak{g}}{\partial p_i} = -\frac{np_i^2}{r^{n+2}}\int_0^r g(\tau,\ \varphi)\,\tau^{n-1}\,d\tau + \frac{p_i^2}{r^2}\,g(p) + \frac{p_i^2}{r^n}\int_0^r \sum_{l=1}^{n-1} \frac{\partial g}{\partial \varphi_l}\frac{\partial \varphi_l}{\partial p_i}\,\tau^{n-1}\,d\tau. \tag{9}$$

Summing equalities (9) over all i = 1,..., n, and recalling that the identity $\sum\limits_{i=1}^{n} \frac{\partial \varphi_k}{\partial p_i} p_i = 0$ holds for spherical angles φ_k, we arrive at (7). It should be noted that if g(p) is bounded, $\mathfrak{g}(p)$ is bounded.

LEMMA 4. Let a function u(x) belonging to $C_2(\overline{\Omega})$ be defined in a bounded domain $\Omega \in E^2$ with a boundary $\partial\Omega$ belonging to class C_2 and let $u|_{\partial\Omega} = \psi(x)$, where $\psi(x) \in C_2(\partial\Omega)$. Let us also assume that we have $\max\limits_{\partial\Omega}|u_x| \leqslant M$. Then, any continuously differentiable function g(p) with $p = (p_1,\ p_2)$ such that g(p) = const for $|p| < 1$, satisfies the inequality

$$\left|\int_\Omega g(u_x)(u_{x_1x_1}u_{x_2x_2} - u_{x_1x_2}^2)\,dx\right| \leqslant C \tag{10}$$

with a constant C which depends only on M, the value of $\max\limits_{|p|\leqslant M}|g(p)|$, the boundary $\partial\Omega$, and the norm of $\psi(x)$ in $C_2(\partial\Omega)$.

Proof. According to Lemma 3, the function $g(u_x)$ can be represented as $g(u_x) = 2\hat{g}(u_x) + \sum_{l=1}^{2} u_{x_l} \frac{\partial \hat{g}(u_x)}{\partial u_{x_l}}$. Let us consider the integral

$$J = \int_\Omega \hat{g}(u_x) u_{x_1x_1} u_{x_2x_2}\, dx. \tag{11}$$

Integrating J by parts twice, we obtain

$$J = -\int_\Omega \left(\hat{g} u_{x_1} u_{x_2x_2x_1} + \frac{\partial \hat{g}}{\partial u_{x_1}} u_{x_1} u_{x_1x_1} u_{x_2x_2} + \frac{\partial \hat{g}}{\partial u_{x_2}} u_{x_1} u_{x_1x_2} u_{x_2x_2} \right) dx + \int_{\partial\Omega} \hat{g} u_{x_1} u_{x_2x_2} \cos nx_1\, ds =$$

$$= -\int_\Omega \left(\frac{\partial \hat{g}}{\partial u_{x_1}} u_{x_1} u_{x_1x_1} u_{x_2x_2} - \frac{\partial g}{\partial u_{x_1}} u_{x_1} u_{x_1x_2}^2 - \hat{g} u_{x_1x_2}^2 \right) dx + \int_{\partial\Omega} \hat{g} (u_{x_1} u_{x_2x_2} \cos nx_1 - u_{x_1x_2} u_{x_1} \cos nx_2)\, ds,$$

from which we have

$$\int_\Omega \left(\hat{g} + u_{x_1} \frac{\partial \hat{g}}{\partial u_{x_1}} \right) (u_{x_1x_1} u_{x_2x_2} - u_{x_1x_2}^2)\, dx = \int_{\partial\Omega} \hat{g} (u_{x_1} u_{x_2x_2} \cos nx_1 - u_{x_1} u_{x_1x_2} \cos nx_2)\, ds. \tag{12}$$

Performing the integration of (11) by parts in another order we also find that

$$\int_\Omega \left(\hat{g} + u_{x_2} \frac{\partial \hat{g}}{\partial u_{x_2}} \right) (u_{x_1x_1} u_{x_2x_2} - u_{x_1x_2}^2)\, dx = \int_{\partial\Omega} \hat{g} (u_{x_2} u_{x_1x_1} \cos nx_2 - u_{x_1x_2} \cos nx_1)\, ds. \tag{13}$$

Adding (12) and (13), we obtain

$$\int_\Omega g(u_x)(u_{x_1x_1} u_{x_2x_2} - u_{x_1x_2}^2)\, dx = \int_\Omega \hat{g}(u_x) (u_{x_kx_k} \cos nx_l - u_{x_lx_k} u_{x_l} \cos nx_k)\, ds = \int_{\partial\Omega} I\, dS. \tag{14}$$

Equality (14) has been derived on the assumption that u(x) is triply continuously differentiable. However, since equality (14) does not contain any third derivatives of u(x), it is clear that it will be correct for u(x) belonging to $C_2(\bar{\Omega})$.

Let us transform I in the same way as was done in [1] (p. 207) by introducing, with the help of an orthogonal transformation, a new coordinate system in the neighborhood of every point x^0 belonging to $\partial\Omega$, the new coordinate system $\{y_1, y_2\}$ being such that the point $y_1 = y_2 = 0$ coincides with x^0, the axis y_1 is directed along the tangent to $\partial\Omega$ at the point x^0, and the axis y_2 is directed outward to $\partial\Omega$ at the point x^0, on the normal running outward from Ω. In addition, let us take a neighborhood of the point x^0 on $\partial\Omega$ such that the equation of the part $\partial\Omega$ of the surface in this neighborhood is of the form $y_2 = \omega(y_1)$. Then, taking into account that $y_k = c_{kl}(x_l - x_l^0)$ and $x_l - x_l^0 = c_{kl} y_k$, where $\cos nx_l = c_{2l}$, $l, k = 1, 2$, we find that at x^0 we have

$$I|_{x^0} = \hat{g}(u_y)(u_{y_1y_1} u_{y_2} - u_{y_1y_2} u_{y_1}) = I_1 + I_2.$$

Let us make use of the boundary condition $u|_{\partial\Omega} = \psi$ for the transformation of I_1. In the neighborhood of x^0, this condition has the form

$$u(y_1, \omega(y_1)) - \psi(y_1, \omega(y_1)) = 0, \tag{15}$$

which is identically fulfilled in y_1 in this neighborhood. Let us differentiate (15) with respect to y_1 and take into account that $\frac{\partial \omega}{\partial y_1} = 0$ at the point x^0. This yields

$$\frac{\partial u}{\partial y_1}=\frac{\partial \psi}{\partial y_1};\ \frac{\partial^2 u}{\partial y_1^2}=-\left(\frac{\partial u}{\partial y_2}-\frac{\partial \psi}{\partial y_2}\right)\frac{\partial^2 \omega}{\partial y_1^2}+\frac{\partial^2 \psi}{\partial y_1^2}$$

and

$$I_1|_{x^0}=-\hat{g}\left[\left(\frac{\partial u}{\partial y_2}-\frac{\partial \psi}{\partial y_2}\right)\frac{\partial^2 \omega}{\partial y_1^2}-\frac{\partial^2 \psi}{\partial y_1^2}\right]\frac{\partial u}{\partial y_2}. \tag{16}$$

It follows from (16) that under the conditions of Lemma 4, the integral $\int\limits_{\partial\Omega} I_1\,ds$ can be estimated in terms of known constants.

Before we examine I_2, let us note that at the point x^0 we have $\frac{\partial}{\partial y_1}=\frac{\partial}{\partial s}$ and $\frac{\partial}{\partial y_2}=\frac{\partial}{\partial n}$, where s is the arc length on $\partial\Omega$ and n the distance along the normal from x^0 to $\partial\Omega$. Therefore, in the neighborhood of the curve $\partial\Omega$ we can use a nondegenerate transformation to introduce new curvilinear coordinates $\{s, n\}$ in which $\int\limits_{\partial\Omega} I_2\,ds$ will appear as

$$\int\limits_{\partial\Omega} I_2\,ds=-\int\limits_{\partial\Omega}\hat{g}\,\frac{\partial^2 u}{\partial s\,\partial n}\cdot\frac{\partial u}{\partial s}\,ds. \tag{17}$$

In order to show that $\int\limits_{\partial\Omega} I_2\,ds$ is bounded, let us integrate by parts the expression

$$-\int\limits_{\partial\Omega}\hat{\hat{g}}(s,\ n,\ u_s,\ u_n)\,u_s u_{sn}\,ds=\int\limits_{\partial\Omega}\left[\left(\hat{\hat{g}}+\frac{\partial\hat{\hat{g}}}{\partial u_s}u_s\right)u_n u_{ss}+\frac{\partial\hat{\hat{g}}}{\partial u_n}u_n u_s u_{ns}+\left(\frac{\partial\hat{\hat{g}}}{\partial s}+\frac{\partial\hat{\hat{g}}}{\partial n}\right)u_s u_n\,ds.\right.$$

This leads to

$$-\int\limits_{\partial\Omega}\left(\hat{\hat{g}}+\frac{\partial\hat{\hat{g}}}{\partial u_n}u_n\right)u_{ns}u_s\,ds=\int\limits_{\partial\Omega}\left[\left(\hat{\hat{g}}+\frac{\partial\hat{\hat{g}}}{\partial u_s}u_s\right)u_n u_{ss}'+\left(\frac{\partial\hat{\hat{g}}}{\partial s}+\frac{\partial\hat{\hat{g}}}{\partial n}\right)u_s u_n\right]ds=\int\limits_{\partial\Omega} I_3\,ds. \tag{18}$$

Assuming that

$$\hat{\hat{g}}=\frac{1}{u_n}\int\limits_0^{u_n}\hat{g}(s,\ n,\ u_s,\ \tau)\,d\tau,$$

we arrive at the equality

$$\int\limits_{\partial\Omega} I_2\,ds=\int\limits_{\partial\Omega} I_3\,ds.$$

Since I_3 can be estimated in terms of known constants, the lemma has been proved.

Remark. If we have $u|_{\partial\Omega}=0$ and $g(p)\geq 0$, then the above requirements concerning the smoothness of $\partial\Omega$ may be relaxed. It is sufficient for $\partial\Omega$ to be a piecewise smooth curve and for the condition $\frac{\partial^2\omega}{\partial y_2^2}\geqslant -M_3$, where $M_3\geq 0$ is a known constant to be satisfied at almost every point x^0 belonging to $\partial\Omega$.

Let us also give a lemma proved in [1] (see p. 94).

LEMMA 5. Let u(x) be a function which is summable in $\Omega \in E^n$ and which for arbitrary $k \geq k_0 > 0$ satisfies the inequalities

$$\int_{A_k} (u-k)\,dx \leqslant \gamma k^{\alpha} \operatorname{mes}^{1+\varepsilon} A_k,$$

where A_k denotes the set of points x belonging to Ω for which we have $u(x) > k$ and γ, α, and ε are constants such that $\varepsilon > 0$ and $0 \leq \alpha \leq 1+\varepsilon$. Then, $\operatorname{vraimax}_{\Omega} u(x)$ does not exceed a constant which only depends on γ, α, ε, k_0, and $\|u\|_{L_1(A_{k_0})}$.

In the following, we will make direct use of a corollary of this lemma to be called Lemma 5'.

LEMMA 5'. If the function u(x), specified in a bounded domain $\Omega \in E^n$, has $\operatorname{vraimax}_{\partial\Omega} u(x)$ bounded and u(x) belongs to $L_l(\Omega)$, whereas $\bar{u}(x) = |u(x)|^{\frac{l}{p}}$ belongs to $W_m^1(\Omega)$, where $1 \leqslant p \leqslant \max\left\{l, \frac{nm}{n-m}\right\}$ with $1 < m \leq n$, and for $k \geqslant k_0 \geqslant \operatorname{vraimax}_{\partial\Omega} u(x)$ the function $v(x) = |u-k|^{l/p}$ satisfies the inequalities

$$\int_{A_k} |\nabla v|^m\,dx \leqslant \gamma k^{\alpha_1} \operatorname{mes}^{1-\frac{m}{n}+\varepsilon_1} A_k \tag{19}$$

with $\varepsilon_1 > 0$, $0 \leqslant \alpha_1 \leqslant \frac{lm}{p} + \varepsilon_1$, then the estimate of $\operatorname{vraimax}_{\Omega} u(x)$ is a constant depending only on γ, α_1, m, n, l, ε_1, and the norm $\|u\|_{L_1(A_{k_0})}$.

To prove this, let us make use of the inclusion theorem for functions belonging to $\overset{0}{W}{}_m^1(\Omega)$, with $m > 1$ (for example, see [1], p. 67)

$$\|u\|_{L_p(\Omega)} \leqslant c\,(\operatorname{mes}\Omega)^{\frac{1}{p}-\frac{n-m}{nm}} \|\nabla u\|_{L_m(\Omega)}, \quad p \leqslant \frac{nm}{n-m},$$

and the Hölder inequality.

Indeed, with the help of the Hölder inequality we obtain

$$\int_{A_k} (u-k)\,dx \leqslant \left[\int_{A_k} (u-k)^l\,dx\right]^{1/l} \operatorname{mes}^{1-1/l} A_k \leqslant \left[\int_{A_k} |v|^p\,dx\right]^{1/l} \operatorname{mes}^{1-1/l} A_k.$$

It follows from the inclusion theorem that we have

$$\left[\int_{A_k} |v|^p\,dx\right]^{1/l} \leqslant c \operatorname{mes}^{\frac{1}{l}-\frac{p(n-m)}{lmn}} A_k \left[\int_{A_k} |\nabla v|^m\,dx\right]^{p/ml},$$

and if inequality (19) is satisfied, then

$$\int_{A_k} (u-k)\,dx \leqslant c_1 k^{\frac{\alpha_1 p}{lm}} \operatorname{mes}^{1+\frac{\varepsilon_1 p}{ml}} A_k.$$

Consequently, the conditions of Lemma 5 are satisfied for $\varepsilon = \frac{\varepsilon_1 p}{lm}$ and $\alpha = \frac{\alpha_1 p}{lm}$.

In order to obtain an estimate of the gradient of the solution of problem (4) inside Ω, we will require additional restrictions of Eq. (1). Let us assume that there exist positive differentiable functions $\varphi_i(p)$ with $i = 1, 2$ such that we have

$$a_{ii}(x, u, p) \leqslant \mu\varphi_i(p), \quad i = 1, 2, \tag{20}$$

where μ is a positive constant, the a_{ii} are the coefficients of Eq. (1), and the $\varphi_i(p)$ are constant for $|p| < 1$.

Let us also strengthen the restriction on the order of growth of the coefficient $a(x, u, p)$ with respect to $|p|$

$$a^2 \leqslant \sum_1^2 \left(\frac{a_{ii}}{\varphi_i}\right)^2 \alpha(|p|)(1+p^2)^2, \quad \alpha(|p|) \downarrow 0 \quad \text{for} \quad |p| \uparrow \infty. \tag{21}$$

Let us show that the following lemma holds.

LEMMA 6. Let $u(x)$ be a solution of problem (4) belonging to $C_2(\bar{\Omega})$, let conditions (2), (20), and (21) be satisfied, together with $\max_{\Omega}|u(x)| + \max_{\partial\Omega}|u_x(x)| \leqslant M$ and let the boundary $\partial\Omega$ together with the boundary values of $u(x)$ be subject to the same conditions as in Lemma 4. Then, $\|u(x)\|_{W_2^2(\Omega)}$ can be estimated in terms of known constants.

Proof. Let us transform Eq. (1) into the form

$$\frac{a_{11}}{a_{12}} u^2_{x_1x_1} - \frac{2a_{12}}{a_{22}} u_{x_1x_1}u_{x_1x_2} - u^2_{x_1x_2} = \frac{a}{a_{22}} u_{x_1x_1} + u^2_{x_1x_2} - u_{x_1x_1}u_{x_2x_2}. \tag{22}$$

Transforming (22) with the help of Cauchy's inequality and taking condition (20) into account, we derive the inequality

$$(u^2_{x_1x_1} + u^2_{x_1x_2}) \leqslant c(\nu, \mu)\left(\frac{a}{a_{22}}\varphi_2\right)^2 + \varphi_2(u^2_{x_1x_2} + u_{x_1x_1}u_{x_2x_2}). \tag{23}$$

Transforming Eq. (1) in another way (by dividing it by a_{11} and multiplying by $u_{x_2x_2}$), we also have

$$(u^2_{x_2x_2} + u^2_{x_1x_2}) \leqslant c(\nu, \mu)\left(\frac{a}{a_{11}}\varphi_1\right)^2 + \varphi_1(u^2_{x_1x_2} - u_{x_1x_1}u_{x_2x_2}). \tag{24}$$

Let us add (22) and (23) and then integrate the result over Ω to obtain

$$\int_\Omega \sum_1^2 u^2_{x_ix_j}\,dx \leqslant c_1(\nu, \mu)\int_\Omega a^2\left[\left(\frac{\varphi_1}{a_{11}}\right)^2 + \left(\frac{\varphi_2}{a_{22}}\right)^2\right]dx + \int_\Omega \sum_1^2 \varphi_i(u^2_{x_1x_2} - u_{x_1x_1}u_{x_2x_2})\,dx.$$

The second integral on the right is finite because of Lemma 4 and, therefore, we have

$$\|u\|_{W_2^2(\Omega)} \leqslant c_1\left\| a\sum_1^2 \frac{\varphi_i}{a_{ii}}\right\|_{L_2(\Omega)} + c_2(\nu, \mu, M, \Omega, \partial\Omega). \tag{25}$$

Let us now make use of condition (21) and the inequality proved on p. 425 of [1]. The fact that the inequality was proved in [1] for functions that are zero on $\partial\Omega$ is unimportant and we obtain

$$\int_\Omega |\nabla u|^4 dx \leqslant c(M, \partial\Omega)\left[\int_\Omega \sum_1^2 u^2_{x_i x_j} dx + 1\right]. \tag{26}$$

Let us subdivide Ω into two parts: a subdomain Ω_1 in which we have $|u_x| \leq K$ and a subdomain Ω_2 given by $\Omega \setminus \Omega_1$. We then obtain

$$\int_\Omega \sum_1^2 \left(\frac{a\varphi_i}{a_{ii}}\right)^2 dx \leqslant c\int_\Omega \alpha(|u_x|)(1+u_x^2)^2 dx \leqslant \int_{\Omega_1} \alpha(|u_x|)(1+u_x^2)^2 dx + \alpha(K)\int_\Omega \sum_1^2 u^2_{x_i x_j} dx + c. \tag{27}$$

Choosing K to be sufficiently large, we can make the coefficient $\alpha(K)$ arbitrarily small and therefore (25), together with (26) and (27), yields

$$\|u\|_{W_2^2(\Omega)} \leqslant c(\nu, \mu, M, \alpha, \Omega, \partial\Omega, \varphi_i). \tag{28}$$

The lemma has been proved.

It should be noted that an estimate for $\int_\Omega |u_x|^s dx$, where s is any finite number greater than unity, follows from (28) by the inclusion theorem.

Now, let us introduce the function $w = (u_{x_1} - k)^2$ where $k > M$, in order to obtain a bound on $\max_\Omega |u_x|$, so that we have

$$w_{x_1} = 2(u_{x_1} - k)u_{x_1}, \quad w_{x_2} = 2(u_{x_1} - k)u_{x_1 x_2}.$$

Let us return to inequality (23). We will investigate it on the set $A_k = \{x\colon u_{x_1} > k\}$. Multiplying (23) by 4w and integrating the result over A_k, we obtain

$$\int_{A_k} \sum_1^2 w^2_{x_i} dx \leqslant c\int_{A_k} \left(\frac{a}{a_{22}}\varphi_2\right)^2 u^2_{x_1} dx + 4\int_{A_k} \varphi_2 (u_{x_1} - k)^2 (u^2_{x_1 x_2} - u_{x_1 x_1} u_{x_2 x_2}) dx = J_1 + J_2.$$

Let us show that we have $J_2 = 0$. Indeed, since w(x) is zero on the boundary of A_k, it can be easily shown by means of an integration by parts that an arbitrary differentiable function $\Phi_2(u_x)$ satisfies the equation

$$\int_{A_k} \Phi_2 (u_{x_1} - k)^2 u_{x_1 x_1} u_{x_2 x_2} dx = -\int_{A_k} \left[\frac{\partial\Phi_2}{\partial u_{x_2}} u_{x_2} (u_{x_1} - k)^2 \; u_{x_1 x_1} u_{x_2 x_2} \left(-\Phi_2 + \frac{\partial \Phi_2}{\partial u_{x_2}} u_{x_2}\right)(u_{x_1} - k)^2 u^2_{x_1 x_2}\right] dx$$

or

$$\int_{A_k} \left(\Phi_2 + \frac{\partial\Phi_2}{\partial u_{x_2}}\right)(u_{x_1} - k)^2 (u_{x_1 x_1} u_{x_2 x_2} - u^2_{x_1 x_2}) dx = 0.$$

If from Φ_2 we obtain φ_2 in the same way as was done in Lemma 3 $(n = 1)$, we obtain the required assertion.

Thus, the following chain of inequalities is valid:

$$\int_{A_k} \sum_1^2 w^2_{x_i} dx \leqslant c\int_{A_k} a^2 \sum_1^2 \left(\frac{\varphi_i}{a_{ii}}\right)^2 u^2_{x_1} dx \leqslant c\int_{A_k} (1 + |u_x|)^6 dx \leqslant$$

$$\leqslant c\,\mathrm{mes}^{1-\frac{1}{l}} A_k\left[\int\limits_{A_k}(1+|u_x|)^{6l}\,dx\right]^{1/l}\leqslant c_1\,\mathrm{mes}^{1-1/l} A_k, \tag{29}$$

where l can be any finite number greater than unity.

Inequalities analogous to (29) can also be derived for $-u_{x_1}$, u_{x_2}, and $-u_{x_2}$ with the help of a suitable choice of w(x) and an estimate of $\max\limits_{\Omega}|u_x|$ follows from them by Lemma 4.

Let us formulate the proposition we have proved in the form of the following lemma.

LEMMA 7. Let u(x) be the solution of problem (4) belonging to $C_2(\bar{\Omega})$, let conditions (2), (3), and (20) be satisfied, and let $\max\limits_{\partial\Omega}|u_x|+\|u\|_{W_2^2(\Omega)}\leqslant M$; then $\max\limits_{\Omega}|u_x|$ can be estimated in terms of known constants.

The following theorems form the principal result of the present article.

THEOREM 1. Let u(x) be the solution of problem (4) belonging to $C_2(\bar{\Omega})$. Let us assume that conditions (2), (20), and (21) are satisfied, that $\psi(x)\in C_2(\partial\Omega)$, that $\partial\Omega$ belongs to C_2 and is strictly convex, and that $\max\limits_{\Omega}|u(x)|\leqslant M$. Then, $\max\limits_{\bar{\Omega}}|u_x|$ can be bounded by a known constant.

THEOREM 2. Under the conditions of Theorem 1, the bound on $\max\limits_{\Omega}|u_x|$ remains valid if the requirement of strict convexity of $\partial\Omega$ is replaced by one of the following:

$$\begin{aligned}&1.\ \frac{\lambda(|p|)}{1+|p|}\leqslant K,\\ &2.\ \frac{\lambda(|p|)}{1+p^2}\leqslant K \text{ and domain } \Omega \text{ is convex (not necessarily strictly convex).}\end{aligned} \tag{30}$$

Remark. If we have $\psi(x) = 0$, the assertion of Theorem 1 remains valid even in the case of nonstrictly convex Ω, and it is possible to relax considerably the requirement that $\partial\Omega$ be smooth (in this connection, see the remark to Lemma 4).

Since Eq. (1) can now be considered as a linear elliptic equation with bounded coefficients,

$$a_{ij}(x, u(x), u_x(x))\,u_{x_ix_j}=a(x, u(x), u_x(x)), \tag{31}$$

$$\nu_1\xi_i^2\leqslant a_{ij}\xi_i\xi_j\leqslant\mu_1\xi_i^2, \tag{32}$$

$$|a_{ij}|+|a|\leqslant\mu_1, \tag{33}$$

where ν_1 and μ_1 are constants defined by Theorems 1 and 2, we can use well-known theorems for linear equations to establish other characteristics of the smoothness of solution u(x) of problem 4. For example, it follows from the theorem proved on p. 267 of [1], that we have $u(x)\in C_{1,\alpha}(\bar{\Omega})$. Let us formulate this theorem in connection with Eq. (31).

THEOREM 3. Let u(x) belong to $W^2_{2,0}(\Omega)$ and satisfy Eq. (31) for which conditions (32) and (33) hold. If $\partial\Omega$ belongs to W_q^2 with q > 2, and both $M=\mathrm{vrai}\max\limits_{\Omega}|u|$ and $M_1=\mathrm{vrai}\max\limits_{\Omega}|u_x|$ are finite, then u belongs to $C_{1,\alpha}(\Omega)$ and the quantity $|u|_{1,\alpha,\Omega}$ is bounded by a constant which depends on ν_1, μ_1, q [defined by conditions (32) and (33)], M, M_1, and, on the boundary, $\partial\Omega$.

Further, if functions a_{ij} (x, u, p) and a(x, u, p) belong to space $C_{0,\alpha}(\mathfrak{M})$, then it follows from Theorems 1, 2, and 3 and the well-known results of Schauder estimates for linear equations that the solution u(x) belongs to class $C_{2,\alpha}(\bar{\Omega})$ when we have $\psi(x) \in C_{2,\alpha}(\partial\Omega)$ and $\partial\Omega \in C_{2,\alpha}$. This allows us to answer the question about the existence of the solution of problem (4) in $C_{2,\alpha}(\Omega)$.

Let us consider the one-parameter family of problems

$$(1-\tau)[a_{ij}(x, u, u_x) u_{x_i x_j} - a(x, u, u_x)] + \tau(\Delta u - u) = 0, \quad u|_{\partial\Omega} = \psi(x), \quad \tau \in [0, 1]. \tag{34}$$

THEOREM 4. Let a_{ij} (x, u, p) and a (x, u, p) belong to class $C_{0,\alpha}(\mathfrak{M})$ and satisfy inequalities (2), (20), and (21). Let us assume, in addition, that we have $\psi(x) \in C_{2,\alpha}(\partial\Omega)$ and $\partial\Omega \in C_{2,\alpha}$. Then, problems (34) are soluble in $C_{2,\alpha}(\bar{\Omega})$ for all $\tau \in [0, 1]$, provided that all of their solutions u(x, τ) are *a priori* bounded by a known constant and that one of the three conditions of Theorems 1 and 2 relating the convexity of Ω to $\lambda(|p|)$ is satisfied.

Theorem 4 is proved on the basis of the *a priori* estimates obtained in Theorems 1 and 2 in the same way as in [1].

In conclusion, the author would like to express her gratitude to N. N. Ural'tseva for her constant attention to the work and valuable advice.

Literature Cited

1. O. A. Ladyzhenskaya and N. N. Ural'tseva, Linear and Quasi-Linear Elliptic Equations [in Russian], Izd. "Nauka," Moscow (1964).
2. S. N. Bernshtein, Collected Works, Vol. III (Partial Differential Equations) [in Russian], Izd. AN SSSR, Moscow (1960).

ON CARLESON'S UNIQUENESS THEOREM FOR ANALYTIC FUNCTIONS WITH A FINITE DIRICHLET INTEGRAL

V. G. Maz'ya and V. P. Khavin

The present article contains the formulation of a theorem which supplements the well-known uniqueness theorem for analytic functions with a finite Dirichlet integral established by L. Carleson [1].

Let G be an open set of space R^n, let e be a compact set contained in G, and let $p \in [1, +\infty)$. Let the symbol U(e) denote the set of all functions φ infinitely differentiable in R^n and with compact carriers in G equal to unity in the set e. The number

$$\inf_{\varphi \in U(e)} \int_G |\operatorname{grad} \varphi|^p \, d\sigma_n,$$

where σ_n is an n-dimensional Lebesgue measure, is called the p-capacity of set e with respect to G and is denoted by $p-\mathrm{cap}_G(e)$*; the p-capacity of set e with respect to R^n will be denoted by p-cap(e). 2-capacity is the well-known Green's capacity in potential theory. We will also require the potential-theory capacity of order α (see [2], p. 33) which we will denote by c_α and the logarithmic capacity $c_{\log}$ (see [3], p. 210).

Let E be a closed set contained in the interval $(-\pi, \pi)$, let $\{(\alpha_\nu, \beta_\nu)\}_{\nu=1}^{\infty}$ be a sequence of nonintersecting intervals whose union contains E, and let $l_\nu = \beta_\nu - \alpha_\nu$. Let E be the intersection of set E and the interval (α_ν, β_ν).

THEOREM 1. *Let us assume that one of the following two conditions is satisfied:*

$$1 < p < 2 \quad \text{and} \quad \sum_{\nu=1}^{\infty} l_\nu \log \frac{l_\nu}{p\text{-cap}\,(E_\nu)} = -\infty,$$

or

$$p = 2 \quad \text{and} \quad \sum_{\nu=1}^{\infty} l_\nu \log \left[l_\nu \log \frac{2 l_\nu}{c_{\log}(E_\nu)} \right] = -\infty.$$

*This notation does not reflect the dependence of p-cap on n. The value of n can always be established from the context.

Let f be a function which is analytic in the unit circle U and which is such that

$$\int_U |f'|^p \, d\sigma_2 < \infty \tag{1}$$

and

$$\lim_{r \to 1} f(re^{i\theta}) = 0 \tag{2}$$

for arbitrary θ belonging to E.

Then, we have $f(z) = 0$ for all z belonging to U.

A function f analytic in U and satisfying condition (1) for p > 2 satisfies the Hölder condition. Therefore, the conclusion of the theorem for p > 2 follows from (1) and (2) and the equality

$$\sum_{\nu=0}^{\infty} l_\nu \log l_\nu = -\infty \tag{3}$$

(see [1]).

L. Carleson [1] has proved the following theorem which we formulate below in the notation of the present article.

THEOREM 2. Let

$$c_\alpha((x-\delta,\ x+\delta) \cap E) > m\delta$$

hold for some $\alpha > 0$ and $m > 0$ and all $x \in E$ and $\delta > 0$ and let condition (3) be satisfied.

Then a function f analytic in U and satisfying condition (1) for p = 2 and condition (2) is identically zero inside U.

It is not difficult to give examples of sets E satisfying condition b) of Theorem 1, but such that we have $c_\alpha(E) = 0$ for any $\alpha > 0$. This means that Theorem 1 allows us to construct "finer" sets of uniqueness than Theorem 2.

The proof of Theorem 1 follows without great difficulty from the estimate of the geometric mean of a function of two variables in terms of the norm of its gradient in L_p (see Lemma 1, below); this estimate is valid for the so-called p-refined functions (see [2]).

Let u be a function defined in an open bounded set G of R^n and let u have generalized first partial derivatives in G with respect to all coordinates p-summable in G [the set of all such functions will be denoted by $L_p^{(1)}(G)$]. A function u is said to be p-refined in G if from any number $\varepsilon > 0$ we can find an open set $g_\varepsilon \subset G$ such that function u is continuous on $G \setminus g_\varepsilon$, and we have $p\text{-cap}_G(g_\varepsilon) < \varepsilon$.

Every function of class $L_p^{(1)}(G)$ coincides almost everywhere in G (with respect to measure σ_n) with a function which is p-refined in G.

Let H be an (n − 1)-dimensional hyperplane in R^n, let E be a compact in R^n, let $\{D\}$ be the sequence of n-dimensional open spheres with centers in H, let $\sigma(D) = \sigma_{n-1}(D \cap H)$, let $c(D) = p\text{-cap}_{2D}(E \cap D)$ where 2D is a sphere concentric with D and with double the radius of D, let $\chi(x)$ be the number of spheres containing the point $x \in R^n$, let $\sigma = \sum_D \sigma(D)$, and let $G = \bigcup D$.

LEMMA 1. Let u be a function belonging to $L_p^{(1)}(G)$, p-refined in G, and equal to zero on E. Then, we have

$$\exp\left[\frac{1}{\sigma}\int_{G\cap H}\log|u|\,d\sigma_{n-1}\right]\leqslant\frac{A}{\sigma}\left\|\chi^{\frac{1}{p}}\operatorname{grad}u\right\|_{L_p^{(1)}(G)}\exp\left(\frac{1}{p\sigma}\sum_D\sigma(D)\log\frac{\sigma(D)}{c(D)}\right),$$

where A is a constant which depends only on p and n.

In order to prove Lemma 1, it is sufficient for us to obtain the upper bound to the integral

$$\int_{D\cap H}\log|u|\,d\sigma_{n-1}$$

in terms of $\|\operatorname{grad}u\|_{L_p(D)}$.

This bound can be easily proved directly, but it is also a particular case of the more general result formulated below, which is of independent interest.

Let U be an n-dimensional open unit sphere, let μ be a nonnegative Borel measure defined on subsets of U, absolutely continuous with respect to p-capacity, and such that $\mu(U) = 1$. Let us define a function γ on the interval (0, 1) as follows:

$$\gamma(t)=\inf_{\{e:\,\mu(e)\geqslant t\}}p\text{-cap}_{2U}(e).$$

LEMMA 2. Let E be a compact subset of U and let u be a function belonging to class $L_p^{(1)}(U)$, p-refined in U, and equal to zero on set E. Then, we have

$$\exp\left(-\int_U\log^+\frac{1}{|u|}\,d\mu\right)\leqslant A\,p\text{-cap}_{2U}^{-\frac{1}{p}}(E)\exp\left[-\frac{1}{p}\int_0^1\log\gamma(t)\,dt\right]\|\operatorname{grad}u\|_{L_p(U)}.$$

Literature Cited

1. L. Carleson, "Sets of uniqueness for functions regular in the unit circle," Acta Mathematica, Vol. 87, Nos. 3-4, pp. 325-345 (1952).
2. J. P. Kahane and R. Salem, Ensembles Parfaits et Séries Trigonométriques, Paris (1966).
3. N. S. Landkof, Fundamentals of Modern Potential Theory [in Russian], GITTL, Moscow (1966).
4. J. Deny and J. L. Lions, "Les espaces du type de Beppo Levi," Annales de l'Institut Fourier, Vol. 5, pp. 305-370 (1953).